CONTINUITY AND CHANGE
IN THE TUNISIAN SAHEL

Continuity and Change in the Tunisian Sahel

RAY HARRIS AND KHALID KOSER
Department of Geography
University College London

Routledge
Taylor & Francis Group

LONDON AND NEW YORK

First published 2004 by Ashgate Publishing

Reissued 2018 by Routledge
2 Park Square, Milton Park, Abingdon, Oxon OX14 4RN
711 Third Avenue, New York, NY 10017, USA

Routledge is an imprint of the Taylor & Francis Group, an informa business

First issued in paperback 2018

A Library of Congress record exists under LC control number: 2003063651

Notice:
Product or corporate names may be trademarks or registered trademarks, and are used only for identification and explanation without intent to infringe.

Publisher's Note
The publisher has gone to great lengths to ensure the quality of this reprint but points out that some imperfections in the original copies may be apparent.

Disclaimer
The publisher has made every effort to trace copyright holders and welcomes correspondence from those they have been unable to contact.

ISBN 13: 978-0-815-38820-3 (hbk)
ISBN 13: 978-1-138-62198-5 (pbk)
ISBN 13: 978-1-351-16112-1 (ebk)

Contents

List of Boxes   *vi*
List of Figures   *vii*
List of Tables   *viii*
Acknowledgements   *ix*

1 Introduction and Overture   1

Part 1 Geographical Contexts   11

2 Physical Environment   13
3 History of Tunisia   25
4 Independent Tunisia   34
5 Economic Development   42
6 Population and Migration in Tunisia   51
7 Women in Tunisia   60
8 Tunisia's Relations with the European Union: History in Five Phases   66
   Alun Jones
9 How 'African' is Tunisia?   74
   Anthony O'Connor

Part 2 Continuity and Change in the Sahel   89

10 Thysdrus and Ancient Rome   91
11 Agriculture and Fishing   98
12 Islam in the Sahel   107
13 The Challenges of the Sahelian Medinas   121
14 Security in Monastir and the Sahel   131
15 The Tunisian Sahel in Context   140

References   *147*
Index   *152*

List of Boxes

2.1   Temperature, precipitation and climate change   19
2.2   The Koran and the physical environment   22
2.3   Environmental Policy Institutions   23
3.1   Hannibal   26
3.2   Ibn Khaldoun   29
3.3   Habib Bourguiba   32
3.4   Celebrating Tunisia's struggle for independence in street names   33
4.1   Remembering Bourguiba in Monastir   38
4.2   Journalism under threat in Tunisia   40
6.1   The demographic transition   54
6.2   Remittances   59
9.1   Economic profile: Tunisia and Niger   79
10.1   Music in the Roman amphitheatre   97
11.1   Agriculture and irrigation around Sahline and Sidi bou Ali   103
11.2   Tunisian government policy on agriculture and fisheries   106
12.1   Christianity and the Koran   111
12.2   Islam, urbanism and architecture   114
14.1   Conservation in Kairouan   134
14.2   Tunisia and the global system   137

List of Figures

1.1	Tunisia on the margins of Europe, the Middle East and Africa	2
1.2	Part of a Tunisian 10 dinar bank note illustrating the significance of The Change of 7 November 1987	4
1.3	The Sahel zone of Tunisia	5
1.4	Field lecture on Roman Tunisia in the amphitheatre in El Djem, November 2002	7
2.1	The major elements of the Atlas mountains in Tunisia	14
2.2	The general circulation of the atmosphere	15
2.3	The role of low pressure cells in producing weather extremes in Tunisia	16
2.4	A NOAA AVHRR satellite image of part of North Africa and Europe, 8 November 1999	17
2.5	Mean annual rainfall (left hand map, units are mm) and mean annual temperature (right hand map, units are °C) for Tunisia, 1901–1960	18
3.1	A Roman mosaic	27
3.2	The Oqba mosque in Kairouan	29
3.3	Habib Bourguiba	32
4.1	The Bourguiba mausoleum in Monastir	34
4.2	Zine el Abidine Ben Ali	39
10.1	Ancient road network of the Thysdrus region	92
10.2	The El Djem amphitheatre today	93
10.3	Sketch of the amphitheatre at its height	94
10.4	A mosaic fishing scene in the Sousse museum	95
10.5	Lines of olive trees	96
11.1	The Nebhana water distribution network	101
11.2	The Sidi bou Ali irrigated perimeter	102
11.3	A broken water pipe losing water in the Bekalta area	102
11.4	Tunisia's national fish exports, 1990–1997	105
12.1	The spread of Islam	108
12.2	Four of the major Islamic scientists of the Middle Ages	113
12.3	The Grand Mosque in Sousse	116
12.4	The ribat in Sousse	117
12.5	The grand mosque in Kairouan	118
12.6	A plan of the grand mosque in Kairouan showing the courtyard, the prayer hall and the minaret	119
13.1	Concentric circle model of a medina	122
13.2	Exterior of a 'typical' medina house	124
13.3	The Monastir medina before and after redevelopment	128
14.1	The medina wall and the gate *Bab Diwan* of Sfax	132
14.2	The ribat of Monastir	133
15.1	A field lecture on security in the ribat in Monastir	143

List of Tables

2.1	Per capita water availability for countries in North Africa in 1990 and projected for 2025	19
5.1	Economic indicators for Tunisia, 1996–2001	43
5.2	Tunisia's main trading partners for exports and imports by value, 1999	44
5.3	Tourism indicators, 1995–1999	46
5.4	Visitors to Tunisia, January to October 2001 and 2002	47
5.5	Investment in agriculture in the Eighth and Ninth Economic Development Plans	48
6.1	Total population (thousands) in Tunisia (1992–2001) and percentage increase on previous year	51
6.2	Projections for Tunisian population (thousands)	52
6.3	The age structure of Tunisia's population (%) (1990–2000)	52
6.4	The largest urban centres in Tunisia	53
6.5	Tunisian citizens in selected EU countries, 2000 or latest available data	56
6.6	Recorded immigration of Tunisian citizens to selected EU countries, 1990–1999	57
6.7	Recorded emigration of Tunisian citizens from selected EU countries, 1990–1999	58
7.1	Summary of the Code of Personal Status	61
7.2	Proportion of women employed in various economic sectors (2001)	62
9.1	Development and welfare indicators 2000–2001	78
9.2	Development and welfare trends	81
9.3	Demographic indicators	82
9.4	International air routes from Tunisia	84
9.5	Countries with embassies in Tunisia (2002)	85
11.1	Tunisian fish catches in tonnes in 1990, 1995 and 1997	105
11.2	The main categories of Tunisian fish exports	106
12.1	The expansion of Islam and the seats of power	109
12.2	Prayer times for Sousse, 1 August 2002	110
12.3	Numbers 1–10 used in the west and in the Arab world	112
15.1	The timetable for field work in the Sousse medina	144
15.2	Questions to guide student information collection in the Sousse medina	144
15.3	Questions to guide the collection of field evidence in rural areas	145

Acknowledgements

The idea for this book arose from a series of undergraduate field courses first at the University of Durham and later at University College London (UCL). At Durham Ray Harris ran the field class jointly with Dick Lawless and Peter Atkins. At UCL we have enjoyed working with three teaching assistants: Charlotte Fry, Gail McLaughlan and Rosie Day. Between us we have taken almost 500 students to the Tunisian Sahel, and this book has benefited enormously from their enthusiasm.

We would also like to thank the Drawing office staff at UCL.

Two chapters in this book are written by guest contributors – Alun Jones and Tony O'Connor (both also at UCL). We are grateful for their input. The chapter by Alun Jones draws on materials from his research project on the European Union and the Mediterranean funded by the Leverhulme Trust.

Chapter 1

Introduction and Overture

Continuity and Change in a Marginal Area

The central theme of this book is continuity and change in the marginal zone of the Sahel of Tunisia. The area has been continuously inhabited for over 3000 years. It was important in Phoenician times because of trade and in Roman times because of food supplies, it was the central area in North Africa during the early days of Islam, it became a protectorate of France in 1881, and it has only recently gained its own independence as a modern state.

The term *Sahel* is more normally associated with West Africa. The term is Arabic and denotes an edge, fringe or coastal plain. In the case of West Africa the Sahel is the edge region or marginal zone south of the Sahara desert. In the case of Tunisia the Sahel is the coastal plain on the eastern side of the country that is itself an edge or fringe. To the south of the Tunisian Sahel is desert and to the north is a wetter climate. To the west lie the Atlas mountains and to the east the Mediterranean Sea. On a grander scale the countries of Tunisia, Algeria and Morocco make up the Maghreb, a region that can be thought of as an island (Rogerson 1998), cut off from the European countries by the Mediterranean Sea and from the countries of Africa by the sand sea of the wilderness of the Sahara. In this Maghreb island Tunisia is on the eastern edge or fringe and the Tunisian Sahel is on the edge of the edge.

Tunisia 'On the Margins'

Context

The physical, terrestrial nature of the marginality of Tunisia and the Sahel is also echoed in other aspects of its geography. The area is 'on the margins' in climate, relationship to Europe, relationship to the Middle East and relationship to Africa (see Figure 1.1).

The rainfall of the Tunisian Sahel is about 300 mm per annum, a figure between the humid temperate agricultural areas to the north and the dry desert to the south. Combined with high summer temperatures, the climate produces an area that is marginal for human habitation. Irrigation is required to maintain agriculture at a sustainable level. On the other hand this climate is a great draw to tourists: sunshine and high summer temperatures are a resource for Tunisian development as significant to the country's economy as (say) oil and gas.

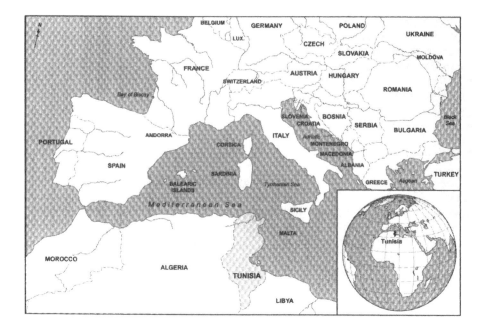

Figure 1.1 Tunisia on the margins of Europe, the Middle East and Africa

Tunisia and Europe

In Phoenician and Roman times Tunisia was not on the margins at all. It was part of the central thoroughfare of the world, the Mediterranean Sea. Transport and communication were commonly by sea and Tunisia was close to the thriving cities of Tyre, Athens and Rome. The Tunisian port city of Carthage was a central node (see Figure 1.1), with excellent communication links by sea to all parts of the Mediterranean.

Yet now Tunisia and the rest of North Africa are regarded as marginal to Europe by most of Europe's population: the countries are foreign, the inhabitants speak Arabic (not a European language) and they follow the non-Christian religion of Islam. Tunisia is an enthusiastic partner with Europe and was the first North African country to sign an association agreement with Europe. But it is not European. One of the criteria for inclusion in the European Union is that a country should be 'European'. Most people in Europe would not regard Tunisia as European, yet it was a French protectorate for 75 years and French is a very common second language, and during Roman times Tunisia (at that time called Ifriqiya, the name later given to the continent of Africa) was arguably more European than the countries of present-day Europe.

Tunisia and the Middle East

In most geography text books on the Middle East (for example, Fisher 1978, Beaumont, Blake and Wagstaff 1988) the Middle East stops at the western border of Libya. In the texts Tunisia is marginal to the Middle East, yet it shares many features in common with the Middle East: language, religion, culture, urban form and market characteristics, including the smells of cardamom, saffron, dried peppers and fennel. A blind woman would have some difficulty discerning her location from the sounds and smells alone whether she was in the souk in Tunis, Cairo, Jerusalem, Kuwait or Muscat.

Despite the strong similarities between Tunisia and the countries of the Middle East _ strongest in terms of religion and language _ Tunisia is frequently also regarded as being marginal to the Middle East. It is in the Maghreb, which is considered somehow different from the Middle East. One could blame the regional geographers of the Second World War and shortly after for this perception. They prepared or were influenced by the Admiralty Handbooks, which separated North Africa from the Middle East. But this does not explain why Egypt usually is considered a Middle East country whereas Tunisia is not. While it is arguable that a definition of the Middle East stretching from Morocco in the west to Iran and Oman in the east is a rather unwieldy region, the similarities of the countries of this region are greater than their differences. Nevertheless in most accounts of the Middle East Tunisia is marginal.

Is this valid? Yasser Arafat was based in Tunis for many years as head of the then Palestine Liberation Organisation (PLO); indeed he was married in Tunis. This provides an illustration of the centrality and marginality of Tunisia at the same time. Arafat was based in an Arab and Islamic country sufficiently friendly under President Habib Bourguiba to allow the PLO to have an operational base, and yet sufficiently distant geographically and emotionally from the land of Palestine to avoid being seen as a threat at that time to the state of Israel.

On the other hand, even in the early years of the 21st century Tunisia looks more to Europe for trade than it does to the countries of the Middle East, despite the similarities with its Arab brothers.

Tunisia and Africa

North Africa is marginal to Africa, and in some minds the term Africa does not include the large and in some cases populous countries of North Africa: Egypt for example has over 60 million people. In European Union policy terms the countries of North Africa do not even contribute to Africa at all: separate policy arrangements are made so that links with North Africa do not count as Africa. In his book *A Passage to Africa* the BBC correspondent George Alagiah (2002) describes many African countries and sees all of them as a disappointment since about 1960. He always means sub-Saharan, Tropical or 'black' Africa when he refers to Africa, never North Africa. Tunisia's population has some genetic contact with black Africa, but the country's gene pool is largely drawn from the Mediterranean.

President Ben Ali is an enthusiast for the African Union, and does put in the hours meeting other African heads of state. But does he really want to align

Tunisia with a failing continent (poverty, disease, malnutrition, HIV-AIDS) rather than with its near neighbour to the north (affluence, high productivity, state welfare, autobahns)? The leader of Tunisia's neighbour, the Libyan president Colonel Qaddafi, is less ambiguous in his keenness to be involved in African affairs, although the actual effects of his enthusiasm are difficult to see in practice. Tunisia is African in a physical sense in that the land is part of the continent of Africa, but it is marginal in most senses that relate to human beings.

Continuity and Change

Continuity in Tunisia is striking: Sousse, Kairouan, Monastir and El Djem all show their past to the visitor in ways that are both brash and subtle. Massive defensive walls stand as proud reminders of earlier isolation, while white tombs and the yellow and blue tiles in mosques tell of their links to Spain and a time when Islam was the religion of Iberia.

Tunisia often picks itself out as a good example of stability and progress in a turbulent world: an identification that is repeated by many world leaders on visits to the country. Its stability is now built on change. The transition from President Habib Bourguiba to President Zine el Abidine ben Ali on 7 November 1987 has been labelled by the government as *The Change*: a change of political regime certainly, but also a change in economic foundations that is captured, for example, on Tunisian bank notes (see figure 1.2). The main sectors of the Tunisian economy, not least the tourism sector, are all characterised by change. Change brought about by internal aspirations such as reduction in poverty and increase in education, and change brought about by external factors such as the events of 11 September 2001.

Figure 1.2   Part of a Tunisian 10 dinar bank note illustrating the significance of The Change of 7 November 1987

Place

The study of Tunisia and of the Tunisian Sahel is therefore the study not just of a particular region but of a wider geography that has links to Europe, to the Middle

East and to a certain extent to Africa. At one level this book is about a specific place: the Sahel of Tunisia (see figure 1.3). At the same time it is about marginal places, central places, places of change, places of tradition, places of modernity, places of continuity. It is about places of study and places of reflection. Places that are little known in the west, or at least little known outside France, and yet places that are familiar in their essence if not in their precise location.

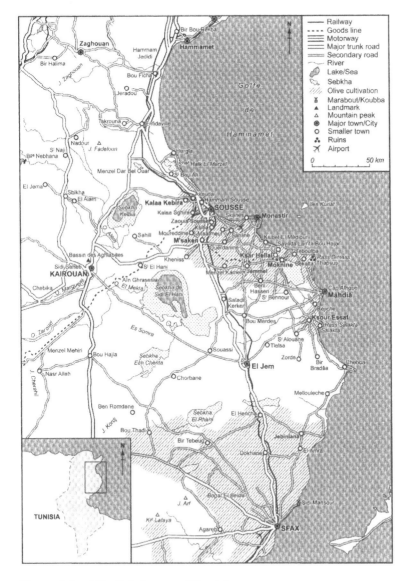

Figure 1.3   The Sahel zone of Tunisia

Field Work in Tunisia

One of the stimuli for this book has been field courses for students of geography at the University of Durham and at University College London (UCL). The University of Durham has had a strong connection with the Middle East for decades and in the 1970s started a series of field classes to the Sahel region of Tunisia. The main purpose of the field class was to provide students with an opportunity to study an Islamic country (and arguably a Middle East country) at first hand. The subsidiary purpose was to give students the opportunity to study a marginal zone as discussed earlier in this chapter.

The Geography Department of UCL used to run field classes to the south of Tunisia to study arid zone geomorphology, and UCL researchers also studied the wetlands of Tunisia as part of the wider study of the hydrology and ecology of wetlands around the world. The Sahel region did not figure in this field work until the 1990s when a new field course was established based in Sousse (see Figure 1.3). The new fieldclass built on the foundations laid at the University of Durham, but made the major focus of study and the sub-themes more explicit, tightened up on the teaching methods and made the field course fit in more simply and more coherently with the rest of the geography degree programme.

The main theme of the Durham and UCL field courses was continuity and change in the marginal zone of the Tunisian Sahel, the main theme also of this book. This central theme has been explored in the field by both Durham and UCL field courses through the following sub-themes.

Islam
Relationship with the global economy
Rural development
Security
Urban form and function.

In the field courses these themes are taught through lectures in the field and in location, for example Islam and the geography of Islam is taught in the grand mosques in Sousse and Kairouan, while Roman Tunisia is taught in the amphitheatre in El Djem (see Figure 1.4). In addition, students carry out field investigations that are either guided around certain questions or are field projects. A collection of such field projects was published by the University of Durham in 1981 (Harris and Lawless 1981) and contained nine reports of field projects carried out by students. In the current book some student projects are reported through the use of boxes to illustrate the more general points in the text.

The field-based study of the Tunisian Sahel provides students with opportunities to study physical and human environments representative more generally: such environments include semi-arid environments, Islam, Arab countries, former French colonies, middle income countries and countries on the margin or periphery.

Figure 1.4 **Field lecture on Roman Tunisia in the amphitheatre in El Djem, November 2002**

Sources of Information on Tunisia and the Sahel

Texts

One of the motivations we had in writing this book is that there is not a large literature on Tunisia in English. Tunisia is the main focus of relatively few academic texts, including: Borowiec (1998) *Modern Tunisia*, Marks and Ford (2001) *Tunisia: Stability and Growth in the New Millennium*, Murphy (1999) *Economic and Political Change in Tunisia: from Bourguiba to Ben Ali*, Perkins (1986) *Tunisia: Crossroads of the Islamic and European Worlds*, Radwan et al. (1991) *Tunisia: Rural Labour and Structural Transformation*, Stone and Simmons (1977) *Change in Tunisia*, Zussman (1992) *Development and Disenchantment in Rural Tunisia* and Zartman (1991) *Tunisia: The Political Economy of Reform*. Because of his importance to the development of the independent state of Tunisia, the first president, Habib Bourguiba, is given close attention by Salem (1984) in *Habib Bourguiba, Islam and the Creation of Tunisia* and by Hopwood (1992) in *Habib Bourguiba of Tunisia*. The literature on the Tunisian Sahel specifically is very small: one contribution is Harris and Lawless (1981) *Field Studies in Tunisia*, a collection of essays on aspects of the geography of the Sahel.

More often Tunisia is set in the context of the Arab world, North Africa, the

Maghreb, the Middle East or the Mediterranean. While Findlay (1994) in *The Arab World* covers a broad scope, there is plenty of material on Tunisia in his book because of his own research in the country. Other texts that include Tunisia in a wider context are Drysdale and Blake (1985) *The Middle East and North Africa*, King et al. (1997) *The Mediterranean. Environment and Society*, Lawless and Findlay (1984) *North Africa*, Long and Reich (2002) *The Government and Politics of the Middle East and North Africa* and Rogerson (1998) *A Traveller's History of North Africa*. While these books provide a spatial context for Tunisia, there is other literature that provides a thematic context. Relevant material on Islam and its Arab and urban context can be found in Al-Bayati (1984) *The City and the Mosque*, Blake and Lawless (1980) *The Changing Middle East City*, Clarke and Bowen-Jones (1981) *Change and Development in the Middle East*, Findlay and Paddison (1984) *Planning the Arab City*, Fishman and Khan (1994) *The Mosque: History, Architecture and Religious Diversity*, Hillenbrand (1994) *Islamic Architecture: Form, Function and Meaning* and Hutt (1977) *Islamic Architecture in North Africa*. Material related to the human context of Tunisia and the country's economic development, including rural development, can be found in Clarke and Noin (1998) *Population and Environment in Arid Regions*, Lawless (1984) *The Middle Eastern Village* and Salem-Murdoch and Horowitz (1990) *Anthropology and Development in North Africa and the Middle East*.

One source of information on Tunisia that is widely used by tourist visitors to Tunisia, but largely neglected in academic texts, is the range of guide books. The most useful one for Tunisia is *The Rough Guide* (Morris and Jacobs 1998): it contains very useful material on history, architecture and Islam in its *Contexts* section and it provides maps and location guides for many of the places discussed in this book, including the towns and cities of the Tunisian Sahel. Other useful guides include the *Insight Guide* (Stannard 1991), the *Travel Guide* (McGuiness 2002) and the *Lonely Planet* guide (Willett 2001).

Tunisia does not seem to be a favourite location for novelists, at least those writing in English. However, a flavour or scent of the Sahel of Tunisia can be experienced in novels set in other parts of North Africa, including *The Sheltering Sky* by Paul Bowles, *L'Etranger (The Outsider)* by Albert Camus and *The Genoa Ferry* by Ronald Harwood. All these novels provide an atmospheric and at times heady mix of the Maghreb, the Mediterranean, Islam and an Arab society.

Web sites

While texts in English about Tunisia and the Tunisian Sahel are not common, there are plenty of web sites that provide news, policy, economic, tourism and factual information about Tunisia. The list below gives suggestions for further exploration, with the proviso that web sites often change, and some go out of existence.

General
Arab Gateway to Tunisia: www.al-bab.com/arab/countries/tunisia.htm
Arabnet and Tunisia: www.arab.net/tunisia/index.html
Government of Tunisia: www.ministeres.tn/
Lexicorient travel guide: lexicorient.com/tunisia/index.htm
Planet Tunisia: www.planet.tn/

Tunisia Online: www.tunisiaonline.com/

Factual information
Arab Data Net: www.arabdatanet.com/, option Tunisia
Canadian International Development Agency: www.acdi-cida.
gc.ca/CIDAWEB/webcountry.nsf/VLUDocEn/Tunisia-Factsataglance
CIA World Factbook: www.cia.gov/cia/publications/factbook/geos/ts.html
Mapquest: www.mapquest.com/atlas/?region=tunisia
Tunisian national statistical agency: www.ins.nat.tn/
United Nations Human Development reports: hdr.undp.org/
United Nations Population Fund: www.unfpa.org/about/index.htm
World Bank reports: econ.worldbank.org/

Media
La Presse newspaper www.lapresse.tn/
Tunisia Daily: www.tunisiadaily.com/
Tunisia Globe: www.tunisiaglobe.com/
Tunisia Online information, including news updates, links to the media, magazines
and online information: www.tunisieinfo.com/
Tunisian radio: www.radiotunis.com/
Tunisian TV: www.tunisiatv.com/

Tourism
Tunisia tourism and travel guide: www.tourismtunisia.com/
Tunisia UK tourist office: www.tourismtunisia.co.uk/

Structure of the Book

This book is divided into two main parts. Part 1 provides the geographical contexts
to the Tunisian Sahel: the subjects examined include physical geography, the
history of Tunisia before and after Independence, economic development,
population and migration, the role of women in Tunisia, and Tunisia's
relationships with Europe, with the Arab world and with Africa. Part 2 focuses on
the Tunisian Sahel zone itself: the chapters consider ancient Rome and Thysdrus
(now El Djem), agriculture and fishing, Islam in the Sahel, the challenges of the
medinas or old cities of the Sahel and security issues in the region. The subjects
covered in each chapter are summarised below.

Chapter 2 examines the basic physical geography of Tunisia and North Africa
and the constraints that the physical environment produces. The chapter covers the
topography, climate, water, soils and vegetation characteristics of Tunisia.

Chapter 3 provides an overview of Tunisian history, from the Phoenicians
through Roman and Arab Rule to French colonisation. Chapter 4 examines post-
Independent Tunisia, comparing the records of its two Presidents.

The focus of Chapter 5 is economic development in Tunisia, particularly the
Ninth and Tenth Economic Development Plans. In addition, the chapter examines
the main features of the Tunisian economy and two of the major sectors, oil and
gas, and tourism.

Chapter 6 is concerned with population and migration in Tunisia. The first part of the chapter analyses trends in Tunisia's population, while the second focuses on international migration and its affects on Tunisia.

Chapter 7 pays particular attention to the situation of women in Tunisia. While gender is a theme that recurs in the analysis throughout the entire book, a chapter on women specifically is justified because of Tunisia's outstanding record on promoting women's rights.

Chapter 8 is written by Alun Jones and examines the relations between Tunisia and the European Union. The discussion is structured into the five major phases of the relationship, starting with Tunisian independence in 1956 right through to the present day and the current Barcelona process.

Chapter 9 is written by Tony O'Connor and asks the question: how 'African' is Tunisia. While Tunisia is a physical part of the African continent, it is arguable that the social, cultural, economic and political links are stronger with Europe and with Arab countries than with Africa. The chapter discusses a range of indicators and dimensions of the question.

Chapter 10 examines the importance of ancient Rome in Tunisia by concentrating on the town of Thysdrus, now El Djem. The remains of a major amphitheatre remind us of the influence of Rome, but so does the extensive cultivation of olives in the Sahel, a landscape of continuity.

Chapter 11 reviews continuity and change in agriculture in Tunisia, and then goes on to examine the fishing industry in Tunisia.

Chapter 12 introduces the religion of Islam, and then focuses on Islam in the Sahel cities of Sousse and Kairouan. Chapter 13 builds on this theme by examining the challenges of the Sahelian medinas. It begins with an explanation of the evolution of medinas, then considers various responses to planning problems.

Chapter 14 looks at the concept of security in the Sahel, and in particular in Monastir. The defensive nature of medina walls is immediately obvious in Sousse, Kairouan, Mahdia and Monastir, but security can also be seen through investments in higher education, the film industry, tourism and the growth of the Internet.

Chapter 15 concludes the book by trying to place the Tunisian Sahel in a wider context. The chapter also looks at how we have taught change and continuity in the Tunisian Sahel to undergraduate geographers in the field.

PART ONE
GEOGRAPHICAL CONTEXTS

Chapter 2

Physical Environment

Introduction

The physical environment of Tunisia provides the canvas or backcloth on which continuity and change in a marginal zone are painted. It is particularly the extremes produced by the physical environment that craft the landscape and give a distinctive character to the human geography of the country. This chapter describes the physical environment of Tunisia, including its topography, climate, water, soils and vegetation. The chapter concludes by reviewing the implications of the physical environment of Tunisia for the Sahel zone.

Topography

Tunisia lies at the eastern end of the Saharan Atlas mountains which extend from Morocco, where the Great Atlas reach to over 2500 m, through the plateau region of Algeria to enter Tunisia at the country's western border. Figure 2.1 shows the three main fingers of the Atlas mountains that enter Tunisia: these three fingers are described below.

The Northern Chain of mountains is a continuation of the Atlas Maritime along the northern coast of Tunisia. The mountains are generally above 800 m and the area has the highest rainfall in Tunisia with over 600 mm per annum. The Dorsale (implying fin or back) has an orientation south west to north east from Kasserine to Cap Bon and the Gulf of Tunis. The mountains of the Dorsale reach over 1000 m in altitude and are an important climatic barrier: in winter they prevent incursions of unsettled weather from the wetter north, while in summer they protect the north from the hot, dry scirocco winds from the Sahara desert. The Southern Dorsale Mountains run west to east through Gafsa to the Gulf of Gabes. They are lower than the Dorsale Mountains, but do have some peaks over 1000 m. They are the northern boundary of the Chott El Djerid, a major inland salt lake or *sebkha* that is dry for most of the year. Together the Southern Dorsale and the Chott El Djerid form a major boundary between the Sahara desert to the south and the rest of Tunisia to the north.

The south of Tunisia, known as the Ksour, is a combination of sand and rock desert. It is largely barren with isolated oases, and has the desert landscapes often termed 'lunar' and made famous in films such as *The English Patient* and *Star Wars*.

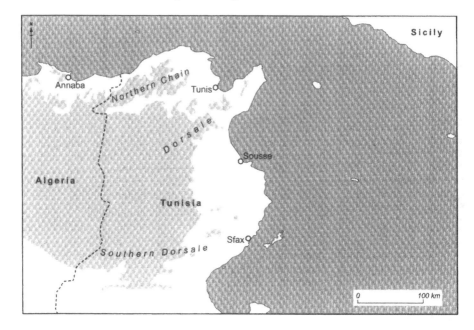

Figure 2.1   The major elements of the Atlas mountains in Tunisia

Climate

Controls on climate

Tunisia is Mediterranean in both location and in climate. The Mediterranean climate is a sub-tropical transition climate, with dry, hot summers and warm/cool, wet winters (Barry and Chorley 1998). The transition in the climate is between sub-tropical high pressure to the south and the westerly airflow and depression tracks to the north. Tunisia's annual climate can be characterised by a transition from a stable summer climate dominated by sub-tropical high pressure to an unstable winter climate characterised by northerly incursions of cool, wet weather.

Figure 2.2 shows schematically the general circulation of the atmosphere. During the summer months of May to October Tunisia falls under the descending limb of the Hadley cell, producing stable weather with clear blue skies, high day-time temperatures (often above 30°C) and low rainfall. The descending limb of the Hadley cell therefore makes a significant contribution to the Tunisian tourist industry by producing hot, sunny weather, although it also produces conditions of high evapotranspiration and extensive water loss.

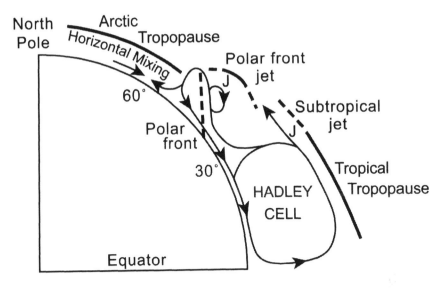

Figure 2.2   The general circulation of the atmosphere
Source: adapted from Barry and Chorley 1998

The period November to April is from time to time influenced by mid-latitude depressions that bring incursions of unsettled weather from Europe. Temperatures fall to around 10°C, often accompanied by strong winds, leaden skies and several days of rain. In the winter the Mediterranean Sea is relatively warm, which enhances depressions because of the land/sea temperature contrast. However, the winter months also often have stable weather conditions with clear blue skies and temperatures of around 15°C, the *winter sun* of holiday brochures.

Low pressure cells

Some of the extreme weather events in Tunisia are caused by low pressure cells migrating through the Mediterranean region. Depending on their position and their track, low pressure cells can bring winds from any compass point, for example southerly winds from the Sahara desert or north easterlies from the Eurasian land mass. Scirocco winds (see Figure 2.3a) develop when a deep low pressure cell travels eastward through the western basin of the Mediterranean and draws in air ahead of it. The winds originate in the Sahara desert and so are hot, dry and very dusty. They can dessicate crops, produce rapid evaporation from open water and make breathing difficult.

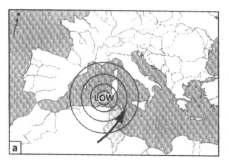 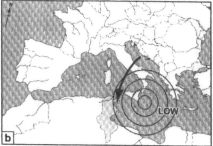

Figure 2.3   The role of low pressure cells in producing weather extremes in Tunisia

By contrast, cold and wet winds from the north and north east can occur where a low pressure cell travels through the eastern basin of the Mediterranean and draws air behind it (see Figure 2.3b). Figure 2.4 shows the swirl of cloud between Italy and Tunisia on a NOAA weather satellite image indicating an active low pressure cell in the central Mediterranean Sea region on 8 November 1999, drawing cold, moist air from eastern Europe to Tunisia. Such air flows can be, for Tunisia, exceptionally cold. In January 1981 a Mediterranean low pressure cell drew in air over Tunisia that had originated in Siberia; this cold air flow brought snow to Sousse for the first time since 1947 and temperatures of less than 5°C for several days.

Regional climates

Rainfall and temperature maps of Tunisia (see Figure 2.5) show a generally west–east orientation of isolines, indicating in turn a north–south change of climate conditions. The Northern Region has rainfall above 400 mm per annum, much of it orographically induced or enhanced by the Northern Chain mountains and the Dorsale Mountains. The Northern Region is the most agriculturally productive region of the country because of its high annual rainfall; the area accounts, for example, for over 70 per cent of Tunisia's cereal production. The Central Region is a triangular shaped area lying between the Dorsale Mountains to the north and the Southern Dorsale to the south. This region, the main focus of this book, has rainfall between 200 and 400 mm per annum, and mean annual temperatures of around 19°C with maxima up to 45°C during the summer. The Southern Region is south of a latitude line through Gafsa with rainfall of only 100 to 200 mm per annum and very high summer temperatures (see Figure 2.5). The region is very thinly populated because of the hot, arid conditions: the oasis of Tozeur for example has a mean July maximum temperature of 39.1°C and a mean July rainfall of zero.

Figure 2.4 **A NOAA AVHRR satellite image of part of North Africa and Europe, 8 November 1999** (15.15 hours, thermal infrared channel 11.5-12.5μm. The swirl of cloud between Italy and Tunisia indicates a low pressure cell, drawing northerly winds from eastern Europe over Tunisia)
Source: Dundee Satellite Receiving Station

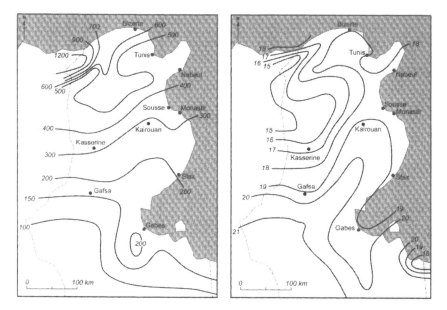

Figure 2.5   Mean annual rainfall (left hand map, units are mm) and mean annual temperature (right hand map, units are °C) for Tunisia, 1901–1960

Rainfall variability and water availability

One characteristic of rainfall in Tunisia is its high variability, which can place great stress on agriculture and on domestic water supplies. The mean annual rainfall in Sousse is 320 mm, but yearly totals have ranged from 179 mm to 525 mm, that is from totals representative of the desert south through to totals representative of the wetter north with its extensive and productive agriculture. Much of the rainfall in the Central and Southern Regions is very intense, and when combined with dry soils can lead to soil erosion, particularly shortly after crops are harvested. The centre and south of Tunisia is therefore at a climatic disadvantage compared to the wetter north, a disadvantage that is likely to increase with predicted climate change during the 21st century.

The results of models of climate change and the work of the Intergovernmental Panel on Climate Change (IPCC 2001) suggest that over the course of the 21st century the climate of Tunisia will become hotter and drier. This will place even greater pressures on agriculture and on domestic water availability in Tunisia. Table 2.1 shows the actual water availability per capita in 1990 and that projected for 2025 for Tunisia and for other North African countries. A report by the US National Research Council (1999) notes that 500 m³ per person per year might suffice in a semi-arid society with sophisticated water management systems: Tunisia was on the border of 500 m³ per person per year in 1990 and will be well below this threshold by the year 2025 if predicted climate change occurs.

**Table 2.1   Per capita water availability for countries in North Africa in 1990
and projected for 2025**

Country	Water availability per capita in 1990 m³/person/year	Projected water availability per capita in 2025 m³/person/year
Algeria	750	380
Libya	160	60
Morocco	1200	680
Tunisia	530	330

Source: Gleick 1992 cited in National Research Council 1999

Box 2.1 Temperature, precipitation and climate change

Based on research work at the Climatic Research Unit of the University of East
Anglia it is possible for the first time to summarise the temperature and
precipitation for Tunisia as a whole. The data in this box were taken from the
Climatic Research Unit (2002) which has created data sets that have been
approved by the Intergovernmental Panel on Climate Change. Meteorological
station data were assimilated on a grid of 0.5° latitude/longitude for the whole
globe and country values extracted from the grid cells that fall within a
country: Mitchell et al (2002) describe the process and the gridded data sets are
described in New et al (1999).

The following diagram shows the annual average temperature for Tunisia for
the period 1901-1998 together with a moving average using an 11 year
window. The mean temperature for the series is 18.8°C. There is a clear trend
of increase in temperature from around 18°C at the start of the 20th century
through to temperatures near 20°C by the end of the 20th century. There has
been a particularly dramatic increase in temperature since about 1980.

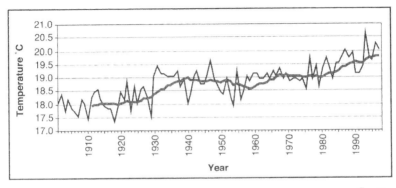

Continued

The average annual precipitation for Tunisia shown in the diagram below reveals a high degree of year-to-year variability. The mean precipitation for the series is 300.4 mm per annum, with a high standard deviation of 61.7 mm and a range from 179 mm (1948) to 525 mm (1977). The high variability of precipitation is evident from the graph, with drought periods in the middle part of the 20th century and in the 1990s.

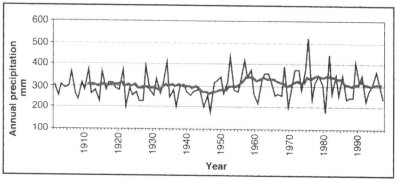

The changes in the Tunisian climate at the end of the 20th century look set to continue into the 21st century. Climate model prediction for the year 2050 from the UK Hadley Centre show for Tunisia:

1　a mean annual temperature increase of 2.0°-4.0°C
2　a reduction in rainfall of 20-30 per cent per annum

Therefore, rainfall is likely to decrease from 300 mm to 250 mm per annum, and its environmental effectiveness will decrease further because of the predicted temperature increase from 20°C to 23°C.

With the low rainfall in the Central and Southern regions and its great variability, drought is an important environmental problem. Since the country's independence in 1956 there have been several serious droughts, most recently the period 1998-2001. These droughts have important consequences as low rainfall means lower crop production, a fall in rural incomes, higher rural unemployment, lower food exports, lower investment in agriculture and greater economic uncertainty. One answer to the problems of low and variable rainfall in Tunisia has been the construction of major dams to collect water and transfer it from the wetter mountain region in the north to the drier central region of the Sahel. This water transfer is used to irrigate new crop areas. One of the first major dam and reservoir investments in Tunisia was the construction of the Nebhana dam which is located 47 km north west of Kairouan in the Dorsale Mountains. The concept of a dam to capitalise on the River Nebhana waters was envisaged at an early stage by the French geographer Despois (1955) before Tunisian independence, and the Nebhana dam, along with its pipeline distribution network, was built from 1962 to 1969 with funding from the World Bank, US and European sources (Davies 1981). Water from the Nebhana reservoir (see Figure 1.3) is used to irrigate over 10,000 ha of agricultural land in the Sahel area: water and agriculture are described in the chapter on agriculture in this book. Dam and reservoir construction has continued in Tunisia since the 1960s: during the years 1997 to 2001 eight new dams and 140 hillside man-made lakes were constructed in Tunisia. The dams and other water schemes now provide water for irrigating 365,000 ha of land in Tunisia.

Soils

The main pattern of soil type distribution in Tunisia follows the rainfall pattern (and the topography pattern to a certain extent) because of the significance of leaching in soil development. In the mountainous regions of the north there is sufficient rainfall to permit some leaching of the soils, leading to the formation of reddish, clay-enriched soils and brown Mediterranean soils. In the Dorsale Mountains there are many calcareous soils, created by the leaching of calcium carbonate into the lower horizons of soil profiles and thereby leading to accumulations of hard calcrete.

In the central zone of Tunisia, including the Sahel, winter rainfall and evapotranspiration are approximately in balance and so leaching of soils is slight. Quaternary alluvial deposits from a wetter period some 5000 years ago together with wind-blown or loess material have produced good environmental conditions for the accumulation of organic matter. In many areas these soils are deep and well drained and have a good agricultural potential as long as water is available. The central part of Tunisia also has highly saline soils in the *sebkhas*, flat areas of clay, silt or sand frequently with saline incrustations on the surface (Brierley 1981, Hollis and Kallel 1986). Sebkhas such as Sebkha Sidi el Hani, Sebkha Kelbia and Sebkha Moknine in the Sahel (see Figure 1.3) are either topographically controlled inland lakes or smaller scale deflation hollows. In all cases they have surface salt crystals caused by salt deposition as ground water is evaporated into the atmosphere. The very high salt concentrations make them devoid of vegetation and they are also a barrier to movement: they can sometimes be areas of 'quicksand' where animals and humans can disappear if trying to walk across their surface.

In the south of Tunisia such sebkhas are very common, the largest of which is the Chott El Djerid, a former lake connected to the Mediterranean Sea but subsequently cut off near Gabes. In the Chott El Djerid region gypsum is abundant and is present in all soils, while 'desert roses' are common features. The dune areas in the south west of Tunisia are raw sands with no soil profile, and the stone or reg desert of south east Tunisia has a desert pavement with no soil horizon development below the surface.

Vegetation

With a Mediterranean climate the vegetation of Tunisia has to withstand long periods of drought and highly variable inputs of rainfall. In addition, the hot summers have high solar radiation amounts that can readily lead to plant dessication. Plants are in competition for water with one another and also with animals and with humans.

Box 2.2 The Koran and the physical environment

The physical environment is an important component of Islam, the main religion of Tunisia. This box presents extracts from the Koran referring to the physical environment, emphasising the importance of the environment in Islam. Source: Bucaille (1976).

The Earth, We spread it out and set thereon mountains standing firm. We caused all kind of things to grow therein in due balance. Therein We have provided you and those you do not supply with means of subsistence and there is not a thing but its stores are with Us. (Surah 15: verses 19-21)

[God] has cast into the ground [mountains] standing firm, so that it does not shake with you. (31: 10)

God is the One Who sends forth the winds which raised up the clouds. He spreads them in the sky as He wills and breaks them into fragments. Then thou seest raindrops issuing from within them. (30: 48)

[God] is the One Who shows you the lightning, with fear and covetousness. He raised up the heavy clouds. The thunder glories His praise and so do the angels for awe. He sends the thunderbolt and strikes with them who He wills while they are disputing about God. (13: 12-13).

Hast thou not seen that God sent water down from the sky and led it through sources into the ground? Then He caused sown fields of different colours to grow. (39: 21)

[God] sends water down from the sky so that the rivers flow according to their measure. The torrent bears away an increasing foam. (13: 17)

[God] is the One Who has let free the two seas, one is agreeable and sweet, the other salty and bitter. He places a barrier between them, a partition that it is forbidden to pass. (25: 53)

Hast thou not seen how thy lord has spread the shade. If He willed, He could have made it stationary. Moreover We made the sun its guide and We withdraw it towards Us easily. (25: 45-46).

The natural vegetation in Mediterranean countries shows certain adaptations to the physical environment. Thick skins and water retaining tissue reduce water losses by evapotranspiration, while thorns discourage animals from grazing. In the extreme environment of a sebkha, where salt concentrations are high, halophytes grow at the sebkha edges in a niche environment, typically visible as a red ground cover.

Cacti, orange trees and olive trees show the major types of vegetation adaptation to the extremes of the climate. Cacti have water retaining tissue and in the case of the commonly-occurring prickly pear cactus sharp thorns to discourage grazing. Oranges have thick, waxy skins that reduce evaporation and protect the inner, water-retaining tissue. Olive trees have small leaves to reduce evaporation and an extensive network of underground roots to capture ground water.

Government Policy, Water and the Environment

Because of Tunisia's climatic marginality water is an important policy issue for the government. The Tunisian government has produced an action plan on water use and on the environment more generally. The plan has the following four major themes.

The effective use or mobilisation of water resources.
The protection of natural resources and the fight against deforestation and soil erosion. This includes afforestation: by 2011 there will be 16 per cent more forests and woodlands in Tunisia.
The avoidance of environmental damage caused by uncontrolled urban expansion. This is a particular problem around the cities of Tunis and Sfax.
The expansion of renewable energy sources. Already 10,000 houses and 200 schools are powered by solar energy, and Tunisia plans to expand this resource.

The rainfall in Tunisia provides 37 billion m^3 of rainwater a year to Tunisia. Of this Tunisia uses 4.08 billion m^3. The main government office for water use development is the National Office for Exploitation and Distribution of Water (SONEDE), supported by the National Office of Sanitation (ONAS). With a budget of 2 billion TD they are making available an extra 1.1 billion m^3 of water through the construction of 21 reservoirs, 200 hillside reservoirs, 1000 hillside lakes, 1850 wells and hydraulic works to distribute water from 2000 wadis. Tunisia is also exploiting alternative water resources such as desalination and treated or recycled water. By the year 2011 the Tunisian government expects to be able to exploit 95 per cent of the water that can be obtained in the country.

Box 2.3 Environmental Policy Institutions

Tunisia has a range of institutions in charge of the environment. They are led by the Ministry of the Environment and Land Use Planning, which is responsible for the national policy on environment protection, and organised through five institutions.

The National Office of Sanitation (ONAS) was established in 1974 to combat pollution and protect water resources. Its task is to contribute to improving the conditions of hygiene and health in urban, tourist and industrial areas.

The National Environment Protection Agency (ANPE) was established in 1988 as a specialised body to control pollution, particularly from industrial sources. It approves the environmental impact studies of major projects.

The National Sustainable Development Commission (CNDD) was established in 1993 within the framework of the follow-up of the Rio de Janeiro summit on sustainable development. Its mission is to ensure the integration of environment issues into development and to promote the approach of sustainable development in all national development plans.

Continued

The Tunis International Center for Environment Technologies (CITET) was established in 1996, also within the framework of the recommendations made by the 1992 Rio conference. CITET promotes environmental learning, knowledge and technologies, working for the transfer of such technologies from the North to the South and adapting them to local conditions.

The National Agency for Coastal Protection (APAL) was established in 1995 following the recommendations of the MED 21 Conference. APAL is responsible for protecting the coastal regions and improving their utilisation and organization, preventing abuses that could harm the coastal environment and conducting studies on the rehabilitation and protection of sensitive and wetland areas.

Implications for the Sahel

The Sahel lies between the Southern Dorsale mountains in the south and the Dorsale Mountains to the north west. The climate of the Sahel area is transitional, with approximately 300 mm rainfall per annum, enough for agriculture when managed carefully. The Sahel is a coastal plain of flat or undulating topography, intersected by shallow wadis only 1-2 m deep (Brierley 1981), and with no permanent river flowing through the region. The Quaternary deposits of water-borne alluvium and wind-blown loess provide areas of relatively fertile soils, although these are counter-balanced by inaccessible and often dangerous sebkhas devoid of vegetation and of no value for agriculture. Soil erosion by hot, dry scirocco winds and by intense rain storms is a problem that diminishes the area of agricultural land and reinforces the requirement to have effective land management schemes. Past climate change has produced a Sahel landscape with relict wadis, while future climate change is likely to put even greater stresses on the Sahel by creating a climate with higher temperatures, lower rainfall and greater rainfall variability in the next 100 years.

The Sahel is an active partner in environmental improvement schemes. In July 2001 the Chamber of Commerce of Ksibet el Mediouni launched a campaign to raise public awareness of the need to conserve water. In 2002 there were *President of the Republic Awards* given to those municipalities and districts that were both clean and active in conservation: in the Sahel the areas given awards were Kairouan, Teboulba and the Kondar district of Sousse.

The physical environment of the Sahel has clear implications for agriculture, but there are also implications for many of the other issues discussed in this book, in particular economic development, tourism, urban growth, environmental security, the relationship of Tunisia with Europe and an understanding of Tunisia since Roman times.

Chapter 3

History of Tunisia

Introduction

In a short chapter it is not possible to do anything other than scratch the surface of Tunisia's dense and complex history, which stretches back over 3000 years to early colonisation by the Phoenicians. The first part of this chapter, covering largely ancient history, focuses particularly on those aspects of Tunisia's history that continue to have an impact on contemporary Tunisia – from the ruins of Phoenician Carthage through Roman artefacts and olive trees to Arab architecture and urbanism. The second part of the chapter examines Tunisia's struggle for independence from French colonisation, a struggle that has defined modern Tunisia. This chapter ends at Independence in 1956, while the next chapter turns attention to the more recent history of independent Tunisia.

Ancient History

The area known today as Tunisia was sparsely inhabited by Berbers when the Phoenicians first landed in the 12th century BC. The subsequent and enduring conflict between the Berbers and the invaders has echoed through Tunisian history ever since: the Berbers similarly tried to resist later invasions by Romans and Arabs, and similarly failed. Today the Berbers comprise a very small proportion of Tunisia's population, and are largely marginalized (Brett and Fentress, 1996).

The Phoenicians

The Phoenicians came from the area of modern Lebanon, and dominated early Mediterranean trade. Their initial interest in North Africa was largely as a staging post for their trade with the metal mines in Southern Spain (Rogerson, 1998). They established a number of ports, first at modern day Utica in 1100BC, and later at the current sites of Sousse and Bizerte. The most famous Phoenician port was Carthage, established in 814 BC by Elissa (also known as Dido), and celebrated in Virgil's Aeneid (Book IV). Beyond the romantic myth, Carthage was an excellent strategic site.

A number of factors encouraged the Phoenicians to develop the hinterland and to establish a greater presence in this area. First, the independence of the Phoenician homeland was under threat by Assyria by the beginning of the 9th century BC. Second, the Phoenicians faced growing competition for Mediterranean trade from the Greeks. Third, there was a growing interest in trading with inland Tunisia, encouraged by inter-marriage with certain tribes of the interior (Rogerson,

1998). As a result, the Phoenicians consolidated their presence in North Africa. The Carthaginian Empire was born and prospered for the next two centuries, developing independently from the Phoenician homeland. The influence of this Empire extended far beyond North Africa – there was even a substantial colony at Pyrgi-Caere just 38 miles from Rome.

From about the 6th century BC onwards the Carthaginians waged intermittent wars with the Greeks, until they suffered a decisive loss in the Battle of Himera (in Sicily) in 400 BC. After Himera they focused even greater attention on expansion within North Africa, colonizing the interior of Tunisia, establishing new cities and exploiting the Berber tribes. During this period, the Romans superseded the Greeks as the principal source of competition for the Phoenicians for control of Mediterranean trade. An earlier alliance with Rome was shattered, and three Punic Wars were fought between the Romans and Carthaginian Phoenicians between 250 BC and 150 BC, including the heroic adventures of the Carthaginian general Hannibal (Box 3.1), and culminating in the sacking of Carthage in 146 BC, which effectively brought to an end the Carthaginian Empire.

Box 3.1 Hannibal

Hannibal was the son of Hamikar Barca, who oversaw the revival of the Carthage Empire after the first Punic War along with his brother-in-law, the Emperor Hasdrubal. Hannibal succeeded to the throne at the age of 25 in 221 BC, after the assassination of his uncle.

Hannibal's conquest of the Greek city of Saguntum on the Catalonian coast, in 218 BC, started the second Punic War. From southern Spain Hannibal launched a pre-emptive strike on Rome. He marched an army of 100,000 men north through Spain and southern Gaul and via the St. Bernard Pass across the Alps, famously along with three elephants.

For 14 years Hannibal waged war on enemy territory. His reputation was as a tactical genius. Fleets from Carthage to the south supported his land assault on Rome from the north. The southern Italian Greek cities of Capua and Syracuse left the Roman fold and entered into alliance with Carthage. The principal flaw in his strategy was to conscript the Gauls of northern Italy, whose ferocity apparently reminded Italians of the need to protect Rome.

After stopping Hannibal in Italy, Rome launched an attack on North Africa – Scipio landed on the Tunisian coast in 204 BC with an army of 30,000 men. Hannibal sued for peace and was appointed a *suffete* (elected judge) in his homeland by the Romans. His enquiry into corruption in Carthage during this period confirmed his status as a hero of the people.

Hannibal took his own life in 182BC in Tyre. Today he is considered a hero across the whole of North Africa.

Roman Africa

Besides the ruins of Carthage itself there are very few material remains of Carthaginian civilization, even though it was renowned for mettalurgy, iron-working, stone-carving, dyeing, joinery and weaving (Rogerson, 1998). This is of course in great contrast to the enormous material legacy of the Romans in modern day Tunisia, from the stunning amphitheatre in El Djem to the mosaics best displayed in the Bardo Museum in Tunis. These are discussed in greater detail in Chapter 10.

Figure 3.1   A Roman mosaic

Initially, however, the Romans, like the Phoenicians before them, had no intentions to colonise the area of Tunisia. They were forced to import soldiers and develop at least a basic infrastructure by a Berber uprising under Jugurtha, which was eventually quelled in 105 BC. Colonisation really took off, however, after Julius Caesar won the Roman Civil War against Pompeii at the Battle of Thapsus (now Ras Dimarse, near Mahdia). Perhaps the most symbolic initiative of Roman colonisation was the refounding of Carthage in 39 BC.

The Roman colony originally comprised two provinces – Ifriqiya Vetus and Ifriqiya Nova – centred in modern Tunisia and extending eastwards into modern Libya and westwards into modern Algeria. In 27BC Emperor Augustus merged these provinces to form the single colony of Ifriqiya Proconsularis. What is striking to note is that the entire African continent thus owes its name to this ancient Roman colony (Ifriqiya having been transliterated to Africa).

For the following two centuries the colony of Africa was one of the most stable parts of the entire Roman Empire. Whereas one small part of France needed four Roman legions (of 6,000 men each) for defence, Africa required only one. This political stability underpinned enormous economic prosperity. Africa became the breadbasket of the Roman Empire, and is estimated to have provided about two thirds of Rome's grain requirements. In the 2nd century AD olive oil production began in earnest in Africa, which further spread prosperity across the Sahel, and, as is explained in Chapter 10, largely funded the construction of the amphitheatre in Thysdrus (El Djem today). In addition to grain and olives, furthermore, Africa was an important source for the Roman Empire of marble, dyes, wild animals, wood and pottery. Rogerson (1998) paints a fairly idyllic picture of daily life in Roman North Africa, highlighting local autonomy, agricultural development and olives, the growth of towns, aqueducts and baths, urban entertainments and gladiatorial games.

A revolt in Thysdrus in 238 marked the end of this golden period. A century of unrest witnessed the unravelling of the Roman Empire, and in Africa Carthage was captured by the Vandals in 439. After a century of Vandal rule, the resurgent Roman Empire, now based in Byzantium (modern Istanbul), made a half-hearted attempt to re-establish the African colony. But within the period of a further hundred years, the area fell to Arab invaders, heralding almost one thousand years of Arab rule.

Arab Rule

The Arab conquest of North Africa began when Alexandria was sacked in 643 and made its presence felt in Tunisia with the Battle of Sbeitla in 647. This invasion was led by Oqba ben Nafi, Commander of the West between 670 and 684 – he also launched his invasion of Spain from this area.

A bewildering succession of Arab dynasties proceeded to rule Tunisia – the Aghlabids (800-909), the Fatimids and Zirids (909-1148), the Almohads and Almoravids (1159-1229), the Hafsids (1207-1574), the Ottamans (1574-1704) and the Husaynids (1704-1881). It is impossible here to do justice to the characters on this stage (such as Abu Yazid – the 'man on the donkey'!), however it is worth singling out the great scholar Ibn Khaldoun, who was born in Tunis in 1332 (Box 3.2) (also see Chapter 11).

Box 3.2 Ibn Khaldoun

Ibn Khaldoun was born in Tunis in 1332 and died in Cairo in 1406. His family was part of the Tunisian elite and he was well educated. He is most renowned as an Islamic historian and philosopher of civilization. During a period of exile in Algeria from 1375 to 1379 he wrote his most influential work, the Foreword (*muqaddama*) which became the first volume of his Universal History (*kitaba-l-abru*). In 1382 he was appointed to a chair at the Al-Azhar University in Cairo where he taught Islamic law. He was famous throughout the Islamic world even during his life, and was received as an honoured guest of the Tatar rule Tamerlane in 1400.

A brief focus on the city of Kairouan provides a useful lens on the major hallmarks of Arab rule in Tunisia. First, Arab Tunisia was inextricably linked with the rapid expansion of the religion of Islam. Kairouan was founded in 670 – according to legend by Oqba – and subsequently became an enormously influential city of pilgrimage. Indeed Islam had significant social impacts across the whole of North Africa, propagated by saints, sufis, scholars and religious notables who quickly spanned out across Tunisia and Algeria (Clancy-Smith, 1994). Kairouan also received the lion's share of Islamic refugees from Spain after Granada fell to the Christian crusades in 1492.

Figure 3.2 The Oqba mosque in Kairouan

Second, urban society flourished under Arab rule. Kairouan's medina is a fine example of the urban structures that came to characterise the landscape of Tunisia and the rest of the Middle East (see Chapter 12). And Kairouan became an important centre of learning and scholarship. Third, and in many ways a manifestation of the interaction between Islam and urban society, Arab rule was characterised by advanced architecture, nowhere better epitomised than in Kairouan's magnificent Oqba mosque (see Chapter 12).

Finally, although French colonisation in Tunisia is officially recorded as beginning in 1881, it was the fall of Kairouan to French troops in October 1882 that really marked the end of Arab rule. Following in the footsteps of the Phoenicians, Romans and Arabs, the French became the latest in a succession of invaders to leave an imprint in Tunisia.

The Struggle for Independence

Since the beginning of the 19th century the French and Italians had been investing heavily in Tunisia. French interests had been growing since their conquest of neighbouring Algeria in 1830. In contrast with Algeria, Tunisia had a stable political system and society during the first half of the 19th century, and presented little threat to European investments there. The Fundamental Pact of 1857 and the Constitution of 1861 represented the introduction – albeit at a superficial level – of liberal principles in Tunisia (Dalacoura, 1998).

French interests increasingly came under threat, however, during the corrupt and incompetent rule of Bey Mohammed as Sadiq (1859-82). He suspended the Constitution in 1864 and actively discredited its liberal principles. He also turned a blind eye to the growth of a nascent Islamic movement that had the purpose of strengthening Muslim civilisation against European encroachment, and dominated more liberal Islamic reformists like Mahmud Qabadu and Khayr al-Din. Frontier clashes with Algeria proved the last straw, and European landowners and holders of now nearly worthless Tunisian government bonds began to clamour for direct intervention by their governments to protect these investments. A new Prime Minister, Kherredine Pasha, steadied the boat, but his dismissal by Sadiq in 1877 effectively sealed Tunisia's fate.

French Colonisation

On March 30 1881 the French army marched into Tunis, virtually unopposed. On May 12 1881 the Bey was forced to sign the Treaty of Bardo (also known as the Treaty of Ksar Said), which gave the French control of foreign affairs and military security. In 1883 the French gained control over Tunisia's domestic affairs too in the Treaty of Marsa. The French army did encounter local pockets of resistance, especially in Kairouan and Sfax, but it was by no means as fierce as Abd al-Krim al-Khattabi's resistance in Morocco, nor that of Emir Abd al-Qadir in Algeria. Although the traditional hierarchy of the Tunisian government was preserved, the Bey was reduced to little more than a figurehead and the French established a parallel 'protectorate' administration under Jules Cambon.

French policy was less brutal and outrageous in Tunisia than it had been in neighbouring Algeria and large-scale colonization never took place. Still it had many of the same demoralising effects on the local population (White *et al.*, 2002). One of the policies with the longest lasting impacts was the establishment of large colonial estates, particularly in the Sahel region. At a stroke tens of thousands of independent farmers were reduced to landless day labourers. One implication was large-scale rural-urban migration, which, as is demonstrated in Chapter 13, was one of the major impetuses for urban change in Sahelian towns.

A second characteristic of colonisation was European immigration. Within twenty years, by the turn of the 20th century, 5 per cent of the population in Tunisia was European. Interestingly, however, the majority of these Europeans were Italians fleeing impoverishment in the *mezzogiorno* of Southern Italy, and not French. Even so, the French numbered about 25,000.

A third characteristic was the establishment of French monopolies. French immigrants almost exclusively filled the higher echelons of the administration. They also dominated the highest paying jobs and the most prestigious sectors of employment. Particularly galling for Tunisians was the fact that under the French their taxes increased significantly, but were spent largely to service the colonizers. Besides politics and economics, a French monopoly was effectively established in the cultural realm too.

Of course France also made some lasting positive contributions to Tunisia's development. Significant improvements were made to the infrastructure in the Sahel – including the construction of wells, roads and dams (Findlay, 1994). And a bilingual Arabic and French education system was instituted. Tunisia's elite acquired bilingual cultural and linguistic skills, reinforced for some with a University education in France. Ironically, these skills were used to great effect in the subsequent struggle for independence.

Nevertheless, resentment against the French protectorate was growing by the beginning of the 20th century and an independence movement began to emerge in Tunisia. As has so often been the case in the history of anti-colonial struggles, the movement was at first splintered and ineffectual. The first flag-bearers of reform were the so-called Young Tunisians who emulated the Young Turks movement of the Ottoman Empire and similarly largely comprised members of the elite and intellectuals. The successor movement, the Destour Party, was founded in 1920, and comprised mainly Tunisian professionals and middle class entrepreneurs. Their rallying call was the 1861 Constitution (*destour* means constitution in Arabic). Like the Young Tunisians, however, the Destour Party could neither speak to nor for the grass roots of Tunisia, and failed to attract widespread support. This was perhaps best exemplified by their decision not to support the *Confederation Générale des Travailleurs Tunisiens* (CGTT), a Tunisian worker's union that was a precursor to the influential *Union Générale des Travailleurs Tunisiens* (UGTT).

Habib Bourguiba and the drive to Independence

In March 1934 a group splintered from the Destour Party to form the Neo-Destour Party. Its leaders were mainly from the Sahel and Djerba, and even though they were also highly educated and members of the liberal professions, their provincial origins allowed them to mobilise Tunisian workers and peasants in a way that had not occurred previously in the nationalist movement (Dalacoura, 1998). Their Secretary-General was Habib Bourguiba (Box 3.3), who was to become the key player in Tunisia's struggle for independence over the next twenty years. Bourguiba quickly emerged as a highly effective populist politician, and attracted widespread support for his calls for self-determination and a return to Islamic culture.

Box 3.3 Habib Bourguiba

Habib Bourguiba was born on 3 August 1903 to middle-class parents in Monastir. He excelled as a student, graduating from the prestigious Sadiki College in Tunis to study law (and marry a French wife) in Paris during the 1920s. What underpinned his later success as an independence leader, however, was his reportedly electric charm and ability to communicate with all walks of life in Tunisia. He became independent Tunisia's first President on 20 March 1956, and was elected President-for-Life in 1974. He was deposed, however, on 7 November 1987, and replaced as President by Zine el Abidine Ben Ali. Despite being deposed, Bourguiba continued to be held in high esteem in Tunisia. He divided his time during his dotage between his presidential palace in Carthage and his hometown Monastir. He died on 6 April 2000 at the age of 97, and his body lies in state in his mausoleum in Monastir.

Figure 3.3   Habib Bourguiba

Momentum towards independence was temporarily halted during the Second World War, remembered most vividly in Tunisia by the exploits of General Montgomery's Eighth Army and symbolised most movingly by a series of Commonwealth War Cemeteries across the country. Its geographical location made

Tunisia a crucial strategic prize in the war, and some 15,000 Allied troops died taking Tunisia from the occupying German army.

After the war, Bourguiba's popular nationalism was augmented by the formation in 1946 of the UGTT, a Tunisian-only trade union federation. Conflict and violence escalated between the French and Tunisians through the 1950s. Bourguiba was repeatedly arrested and exiled, and in December 1952 Farhat Hached, secretary-general of the UGTT, was assassinated in Tunis by French settlers. His name, like those of many other independence leaders, is still commonly evoked across Tunisia (Box 3.4).

Box 3.4 Celebrating Tunisia's struggle for independence in street names

Many of the key figures and even dates in recent Tunisian history are remembered in street names. Every town in Tunisia will have at least one street named for Habib Bourguiba, and other leaders commonly commemorated are Farhat Hached and Habib Achour, both Trade Union leaders and key players in Tunisia's struggle for independence. Three dates in particular also recur in street names. 9 April (or Martyr's Day) was the day in 1938 when French troops shot dead 122 Tunisians. 20 March 1956 was the day Tunisia won her Independence from France. 7 November 1987 (or New Era Day) was the day Habib Bourguiba was deposed as President by Zine el Abidine Ben Ali.

Obstinacy on the part of the French eventually gave way to concession, as they suffered colonial setbacks in Indochina and as the winds of change across the world – in India and elsewhere in Africa – began to blow inevitably towards the end of colonisation. A protracted series of French-Tunisian talks ended with agreement for the independence of Tunisia in June 1955, when Bourguiba returned from exile to an ecstatic welcome in Tunis. There was some opposition to Bourguiba from within the UGTT and the Neo-Destour Party (notably by Salih ibn Youssef) for his decision initially to accept France's offer of 'autonomy' rather than full independence. Nevertheless, on March 20 1956 Tunisia finally did achieve full independence, and Habib Bourguiba was appointed the first president of the Republic of Tunisia, thus joining the ranks of the new political elite of the Arab World (Milton-Edwards, 2000).

Conclusion

History is alive in Tunisia. It is perfectly possible to sit in sight of the ruins of Carthage, while ordering olives (in French) and listening to a muezzin call the faithful to prayer, in a café on Avenue Habib Bourguiba. As the following chapters demonstrate, historical antecedents run strongly though almost all aspects of Tunisia – from politics through economics to society and culture. History provides the continuity over which change has been laid, but rarely in Tunisia has history been totally eradicated by change.

Chapter 4

Independent Tunisia

Introduction

Since Independence there have been only two Presidents of Tunisia – Habib Bourguiba (1956-87) and Zine el Abidine Ben Ali (1987-present). Even though the term *'le changement'* ('the change') is widely used in Tunisia to describe Ben Ali's succession, and even though Ben Ali did introduce some radical departures in policy, there has also been much continuity between the two Presidents. Both, for example, have been challenged by contradictions surrounding Islam and modernization, security and human rights, and political liberalism and control. This theme of continuity and change is highlighted in this chapter by comparing and contrasting in turn three aspects of each man's presidency: their impacts on Tunisian society, politics and the economy.

Habib Bourguiba

Figure 4.1   The Bourguiba mausoleum in Monastir

Engraved on the door of Habib Bourguiba's mausoleum in Monastir is the epithet 'Liberator of women, builder of modern Tunisia'. We shall see in Chapter 7 that the first half of this couplet is well deserved – there is no doubt that women enjoy greater rights in Tunisia than in any other country in the Maghreb or Middle East. To what extent, however, can Bourguiba also be described as the 'builder of modern Tunisia'? This question can be answered by examining his record on social, political and economic reform.

Society

Bourguiba is most renowned for his Code of Personal Status, which took significant steps towards achieving equal rights for women (see Chapter 7). This can be viewed as one objective in a wider aim to institutionalise a liberal version of Islam and introduce a broadly secular society in Tunisia. Other reforms were targeted quite specifically on weakening the religious establishment. The tradition of *habous* (religious endowments) was abolished, legal and educational systems were unified, and Zaytouna – the ancient Islamic university and religious centre in Tunis – was downgraded to a theological faculty of the modern university. Tunisia kept the French weekend of Saturday and Sunday, in contrast, for example, with Algeria where Friday (the Muslim day of prayer) became the end of the week. Furthermore, in 1959 Bourguiba introduced a new constitution. This guaranteed full civil, political and social rights for all Tunisian citizens regardless of their race, religion and sex.

Bourguiba even attempted to put an end to the tradition of the Ramadan fast. Fasting – one of the five pillars of Islam (see Chapter 12) – is excused for certain categories of people (for example pregnant women) and also in certain circumstances, including during a holy war or *jihad*. Bourguiba argued that all Tunisians were engaged in a *jihad* against poverty and underdevelopment, and that on these grounds should not have to fast. In this instance opposition centred on clerics in the religious city of Kairouan thwarted Bourguiba. A smaller victory was, however, won when Bourguiba ruled that his citizens could eat in public during Ramadan should they wish to.

Judged on social reform, it would be hard to deny that Bourguiba did indeed lay the foundations that make Tunisia the most secular Muslim state in the world today. Such was his tireless pursuit of secularism that it is reported that in his early years as President, whenever a new committee was formed, Bourguiba would ask: 'Does it have a Jew and a woman?' (Field, 1994: 277). As we shall see below, however, social reform was not without its opponents.

Politics

What cannot be avoided is that Bourguiba's largely admirable social reforms were only possible because of his rather less admirable and at times downright ruthless domination of Tunisian politics.

White *et al.* (2002) identify a number of distinct phases in the evolution of the political system under Bourguiba. These are only briefly summarised here. Between 1955-59 Bourguiba consolidated his power. In April 1956 he became Prime Minister (since March he had, of course, also been President) and appointed a cabinet in which 16 of 17 members were from the Neo-Destour Party. In July 1957 he abolished the position of Bey, thus reinforcing his own position as Head of State. Bourguiba held his first national election on 8 November 1959 in which he was elected unopposed and all 90 candidates for the Chamber of Deputies (*Majlis al-Nawaab*) had Neo-Destour backing.

The second phase, between 1960-64, was characterised by a series of internal and external challenges. Skirmishes broke out between Tunisian and French troops in July 1961 when Tunisia demanded that France evacuate their large naval base in Bizerte. An assassination plot against Bourguiba was uncovered in December 1962, which provided a pretext to imprison or execute several military officers and to ban the Communist Party in January 1963. Also of note during this period is that in 1961 the Neo-Destour Party was renamed as the Destourian Socialist Party (PSD). And Salih ibn Youssef – who had opposed Bourguiba's initial agreement for autonomy from the French in 1955 – was assassinated in France by Tunisian agents.

The period 1964-1969 was dominated by the efforts of Ben Salah – another former rival – as Minister for Planning and National Economy. His success threatened Bourguiba, who by the end of the 1960s was already chronically ill and often abroad for treatment, and in September 1969 Ben Salah was dismissed. He was later arrested and sentenced to ten years hard labour (he escaped to France after three years).

Bourguiba's authoritarian regime was attracting increasing opposition, and during the early 1970s he responded by swinging to political liberalism. Several former officials, for example, were reappointed – including Habib Achour as head of the UGTT. It soon became apparent, however, that these were largely cosmetic measures, and at the 1974 PSD Congress Bourguiba appointed himself as President-for-Life of Tunisia. For the next ten years there was effectively no political reform in Tunisia. The PSD became increasingly monolithic and lost its earlier political effectiveness. Tunisian politics became conflict-ridden, and Bourguiba struggled to contain new political opposition.

The Economy

Like his political approach, Bourguiba's economic policies were also largely reactive. Neither displayed the vision of his proactive social reforms. From 1956 to 1961 Tunisia pursued a liberal, *laissez-faire* economic policy. Returns on investment were, however, disappointing, at least in part because of the departure of French and Italian colonial communities and of the skilled indigenous Jewish population (White *et al.*, 2002).

In response, Bourguiba turned to Ben Salah in 1961 to develop a planned economy. A ten-year economic development strategy was drafted (*Perspectives décennales de développement*) – and the five-year plans described in detail in the next chapter were instigated. The aims of the first two plans were fourfold: decolonisation; reform of economic structures; human development, and internal investment. Alongside economic reform there was a significant campaign of agricultural collectivisation into cooperatives.

By 1969 this state-led economic experiment had largely failed. State-run enterprises had low productivity and collectivisation was unpopular with the rural middle classes. Interestingly, state socialism had also lost its popularity after the discrediting of Egyptian President Nasser's 'Arab socialism' following his humiliation during the 1967 Arab-Israeli war.

At the beginning of the 1970s, therefore, Tunisia returned to a free-market economy. A policy of economic openness – *infitah* – was formalised. According to Martin-Edwards (2000), *infitah* was a precursor for President Gorbachev's *glasnost* in the Soviet Union almost 20 years later. Foreign investment was actively encouraged in Tunisia, through financial incentives and an emphasis on cheap and largely non-unionised labour. Another strand of *infitah* was the promotion of tourism. During the late 1970s and early 1980s these policies had dramatic success – between 1973 and 1980 gross domestic production (GDP) grew by at least seven per cent annually. As we shall see in the following section, however, the boom came to an end by the middle of the 1980s.

'Le Changement'

Bourguiba may have been the 'builder of modern Tunisia', but the foundations he laid also contained the seeds of his own destruction. By the middle of the 1980s Tunisia was rapidly unravelling around Bourguiba, undermining his power and position.

At least in part in response to his social reforms, an Islamist movement emerged in Tunisia during the early 1970s. It combined several groups including some radical offshoots, but the most influential became the *Mouvement de la Tendence Islamique* (MTI) – which adopted this name in 1979 (in 1989 it was renamed *An-Nahda*). The MTI was associated with a wider Islamist movement across the Arab World at this time whose members were known as the *intégristes*. The MTI rejected all notions of democracy, human rights and equality of sexes as Western imports and as contrary to the Koran and Islamic law (Dalacoura, 1988).

Alone the Islamist movement might not have caused major problems for Bourguiba, but its evolution also coincided with the emergence of a labour movement and increasing opposition to the excesses of his political regime. Of enormous symbolic significance, the UGTT held Tunisia's first national strike in 1978. On 26 January – a day that became known as 'Black Thursday' – the strike was violently put down: over 100 strikers were killed and the UGTT leader Habib Achour was imprisoned. Ironically, this crackdown on the UGTT was the catalyst for the politicisation of the Islamist movement. Under the leadership of Rachid Ghannouchi the MTI became increasingly popular, and increasingly dangerous – in 1986 tourist resorts in Monastir were bombed in an effort to destabilise the economy.

Opposition to social and political reform combined fatally with the end of the economic boom. As we shall see in Chapter 8, European Community (EC) member states began to put up tariffs against cheap Tunisian industrial imports; plummeting oil prices flooded the labour market with Tunisian workers returning from the Gulf states, and a series of droughts in the early 1980s drastically reduced agricultural productivity (Zussman, 1992). Unemployment spiralled to 20 per cent, foreign debt rocketed to nearly $5 billion and foreign exchange reserves shrank to almost zero.

In 1983 the government implemented a series of drastic austerity measures, encouraged by the International Monetary Fund. These included the removal, on December 29, of the subsidy on bread, resulting in an overnight increase in bread prices. The so-called 'bread riots' ensued on January 5 1984. These were immediately and viciously suppressed and at least 150 people were killed. Bread subsidies were restored on January 6, but the damage had clearly been done.

The suppression of the 'bread riots' was orchestrated by the Minster of Interior – Zine el Abidine Ben Ali, and established his reputation. In October 1984 he became the first military officer in post-Independence Tunisia to hold a high political post when he was appointed Secretary of State for National Security.

All of these factors – social discontent, political opposition, economic deterioration and the ascendancy of the powerful Ben Ali – coincided with Bourguiba's increasing frailty. Not only had he been chronically ill since the late 1960s, he was also by now displaying signs of senility. On November 7 1987 Ben Ali had seven doctors write independent reports on the President, and all of them agreed that he was no longer fit to serve. The President-for-Life was removed in a bloodless coup on the same day, marking *'le changement'* in independent Tunisia. Bourguiba continued to live in Tunisia and died on April 6 2000 (Box 4.1).

Box 4.1 Remembering Bourguiba in Monastir

Habib Bourguiba died on April 6 2000 at the age of 97. His coffin is on public view in his mausoleum in his hometown of Monastir. Monastir today is infused with the memory of Bourguiba. It was his birthplace, and his childhood home has been preserved, even though it attracts little attention from tourists as it lies away from the centre of town. His school days are commemorated with a golden statue of the boy Bourguiba, books in hand. Monastir has an annual festival on Bourguiba's birthday, is the location of one of his Presidential palaces, and is also the site of the modern Bourguiba mosque, reputed to accommodate one million worshippers. Most poignantly, Monastir is the site of Bourguiba's mausoleum, built in the late 1960s over 30 years before his eventual demise! As well as his coffin, the mausoleum also contains a mock-up of his office as President, containing gifts from various world leaders.

Zine el Abidine Ben Ali

Far less has been written about the presidency of Ben Ali than that of Bourguiba, and so this section is necessarily brief. One reason is that until only recently, when his human rights record began to attract international criticism, Ben Ali's Tunisia has been fairly stable and largely un-noteworthy in international affairs. Certainly compared with its neighours Libya and Algeria, Tunisia has rarely entered the headlines anywhere. One exception was the bombing of a synagogue in Djerba in 2002, reportedly by Al Qaeda sympathizers, but this appears to have been a one-off tragedy rather than the symptom of a new Islamic fundamentalism in Tunisia.

Figure 4.2   Zine el Abidine Ben Ali

Society

As is explained in Chapter 7, Ben Ali amended (positively) the Code of Personal Status in 1992, thus building on Bourguiba's legacy and reputation for social reform. The standout issue in recent years, however, has been his record on human rights.

A recent report by Amnesty International (2003) has criticised the Ben Ali regime for abusing human rights in a range of ways. At an individual level this includes: arbitrary arrests, torture, no – or limited – access to lawyers, and tampering with the files of defendants. At a broader level the regime is criticised for its lack of democratic accountability, for politicising the judiciary, and for clamping down on the freedom of speech. Particular criticism has been reserved for the treatment of journalists who are critical of the government (Box 4.2). Such abuses notwithstanding, most commentators still agree that civil society remains stronger in Tunisia than in most other Arab states today (Milton-Edwards, 2000).

Politics

Most of Ben Ali's excesses have been targeted on political opponents – who appear to include even those who simply criticise the regime – and in particular on the Islamist movement. This contrasts with his approach when he first became President. The catalyst for overthrowing Bourguiba had been the conflict between the latter and Rachid Ghannouchi (leader of the MTI), who had been condemned to life with hard labour in 1987. Ben Ali realised this would only make a martyr of Ghannouchi and his cause, and thus freed him as soon as he became President. In the following months a further 6,000 political prisoners were also released. Leaders of the Islamist movement and other political opponents who had fled Tunisia were invited back (including Ben Salah who returned in 1989). Publication of newspapers that had been banned was allowed again and the press code reformed to give them greater freedom of expression. In April 1988, furthermore, a law was passed to legalize political parties in addition to Ben Ali's (which he renamed the *Rassemblement Constitutionel Democratique* – RCD). The MTI subsequently transformed into a political party, the *Hizb An-Nahda* (Renaissance Party). (The English transliteration of the party's name also appears sometimes as *Ennahda*).

Box 4.2 Journalism under threat in Tunisia

In 2001 Amnesty International presented a 'Special Award for Human Rights Journalism under Threat' to Tunisian journalist Sihem Ben Sedrine. In April 2000 she had been arrested and beaten when she attempted to visit another journalist, Taoufik Ben Brik, who was on hunger strike in protest at police harassment. Her on-line newspaper, *Kalima*, cannot be printed and sold in Tunisia as it is not authorised by the government. On 26 June 2001 she was arrested and imprisoned for almost seven weeks shortly after returning from a visit to Europe where she had spoken out against the Ben Ali government.

Things turned sour with the April 1989 general election. Six opposition parties – and *An-Nahda* – fielded candidates (*An-Nahda* was not permitted to enter the election as a political party, but fielded independent candidates). Although the six fared poorly (winning about one per cent of the vote), most commentators agree that the Islamists won about 18 per cent of the vote. Official results, however, showed Ben Ali winning 99 per cent of the vote, and RCD won all the seats in the *Majlis* (Chamber of Deputies). In part this outcome was the result of Tunisia's 'first past the post' electoral system; there were however also allegations of widespread irregularities.

The Islamists considered themselves duped. *An-Nahda* turned against the entire system and its members became increasingly vociferous in their opposition to Ben Ali. His response was in turn reactive – leading Islamists were imprisoned and in 1989 the theological faculty of Zaytouna was closed down altogether. In May 1991, Tunisian police uncovered a plot to overthrow the regime. Caches of weapons were found on university campuses in Tunis and Sfax, with lists of people to be assassinated and a plan for a missile attack on the President's aeroplane.

Since the beginning of the 1990s Ben Ali has been consolidating power in the face of a continuing Islamist challenge, while simultaneously trying to maintain the image of a liberal, democratising state – in particular in order to continue attracting foreign investment. This image increasingly appears to belie the truth. The critiques of his human rights record cast a shadow over the idea that Tunisia really is a liberal state. Similar doubt exists over the claim that Tunisia is a democratic state. Even though Ben Ali has publicly committed himself to multi-party elections, effectively they have not yet occurred (Gerner and Schrodt, 2000). There does exist a multi party political system – although *An-Nahda* is excluded – but Ben Ali has consistently won over 99 per cent of the vote in general elections (1994 and 1999). Nevertheless, even critics reckon Tunisia still offers the greatest hope eventually for genuine political liberalization in the Arab World (Milton-Edwards, 2000).

The Economy

As we shall see in the next chapter, Ben Ali has a good economic record (White *et al.*, 2002). While the economy ailed through the period of the seventh development plan (1987-1991), ever since it has been growing. Light manufacturing and the service sectors have expanded significantly. Greater emphasis has been placed on mineral extraction – and special attention has been paid to the natural gas sector after the Miskar natural gas field came onstream in 1995 (British Gas is an important investor). And tourism has become an economic mainstay.

Ben Ali has also committed significant resources to social spending. There is a good social security system, transfers and subsidies assist the poor, and there is free health care and schooling in Tunisia. This is the best record of any country in the Middle East or North Africa (Moghadam, 2000).

Tunisia's economy does nevertheless also face considerable challenges. In 2000 seven per cent of the population was still estimated to be below the poverty line (Moghadam, 2000). Tourism has yet to recover from the downturn following the events of September 11th 2001. The economy is heavily in debt – in 1999 Tunisia owed $11 billion. And the agricultural sector is struggling (White *et al.*, 2002).

Conclusion

To what extent did *'le changement'* actually represent a watershed in the recent history of independent Tunisia? Arguments can be mobilised in both directions. Ben Ali appears not to share Bourguiba's particular zeal for social reform. He has been – at least on the surface – far more tolerant of political reform. And he has overseen the substantial recovery of Tunisia's economy.

On the other hand: there has been a tendency to 'celebrate' Bourguiba rather than criticise him – to emphasise his Code of Personal Status, for example, while downplaying the fact that he had political opponents imprisoned and sometimes murdered. And has Ben Ali really been more open to political reform? Admittedly he has not (yet) appointed himself as President-for-Life, but there have to be questions asked when he consistently wins 99 per cent of the vote in national elections. Finally, while the economy has indeed recovered under Ben Ali, he has nevertheless pursued the strategic five-year development plans introduced by Ben Salah under Bourguiba.

It is tempting to conclude that *'le changement'* was more symbolic than real. And such a conclusion would of course also reinforce the idea of Bourguiba as the 'builder of modern Tunisia'. He lay the foundations for the most secular Islamic state in the world. He also lay the foundations for a political system that tolerates human rights abuses.

Chapter 5

Economic Development

Introduction

Since independence in 1956 Tunisia has pursued a series of five-year economic development plans. It is one of the few countries that maintains the momentum of using five-year plans on a national basis for economic and social planning. This chapter discusses the Ninth Economic Development Plan (1997-2001) and the Tenth Economic Development Plan (2002-2006). These development plans provide a framework for national economic and social development and in particular provide the objectives for growth and development. Before discussing these two plan periods this chapter also examines the main features of the Tunisian economy (production, trade and foreign investment) and two of the major economic sectors – oil and gas, and tourism. The chapter concludes by reviewing economic development in the Sahel region.

Main Features of the Tunisian Economy

Production

As an indicator of the size of the economy, the Gross Domestic Product (GDP) of Tunisia in 2001 was 29.4 billion TD (Tunisia Online 2003). The exchange rate is about 2TD = £1 and 1.3TD = US$1. For comparison with other countries, the Tunisian GDP represents a purchasing power parity of US$64.5 billion. Growth of national GDP in recent years has been at a rate of over 5 per cent per annum. Table 5.1 shows economic indicator data for Tunisia for the period 1996-2001. All the indicators show improvement in the economy, with GDP per capita rising above 3000 TD for the first time in 2001. The level of savings and investments is about one quarter of GDP and is growing, while inflation has declined from 3.7 per cent in 1996 to 2.9 per cent in the year 2000.

The largest sectors of the Tunisian economy are agriculture, mining, energy, manufacturing and tourism. In 2001 services accounted for 59 per cent of GDP, industry 29 per cent and agriculture 12 per cent (World Bank 2002a). Average annual growth rates in the industrial and service sectors have increased from 3 per cent per annum in the 1980s to 6-7 per cent in 2001, but the annual average growth rate in the agriculture sector has fallen from 4 per cent per annum to -1.0 per cent in 2001 (World Bank 2002b). Agriculture is a problem in an otherwise buoyant economy because of a reliance on traditional products (such as olives, olive oil, grain, dairy products and dates), competition from Mediterranean countries, adverse trading conditions and an extended drought.

Table 5.1   Economic indicators for Tunisia, 1996–2001

Indicator	1996	1997	1998	1999	2000	2001
Gross Domestic Product (million TD)	19066	20989	22581	24672	26923	29423
Gross Domestic Product per capita (TD)	2097	2268	2419	2609	2815	3042
Gross National Product (million TD)	18056	19896	21639	23657	25831	28239
National savings (per cent of GDP)	23.7	24.5	24.5	25.1	26.2	26.4
Investment rate (per cent of GDP)	23.2	24.7	24.9	25.4	26.2	26.4
Rate of inflation (per cent)	3.7	3.7	3.1	2.7	2.9	

Source: INS 2002

International trade

Tunisia's exports are orientated primarily to western Europe, particularly to France, Italy and Germany (CIA 2002). The main export commodities are agricultural products, phosphates, chemicals and hydrocarbons, mechanical goods and textiles. The main imports are food, chemicals, hydrocarbons, machinery and equipment, and they originate largely in the same partner countries, namely France, Italy and Germany. Table 5.2 lists the top 10 trade partners with Tunisia in 1999 for both exports and imports.

One surprising element of the international trade data is the limited presence of other Arab or Middle East states: Table 4.2 shows only Libya and Turkey in the top 10 of exporting partners and only Libya in the top 10 of importing partners. Algeria and Morocco together take only 1.3 per cent of Tunisia's exports and do not appear at all in the top 20 countries for imports to Tunisia. There is some evidence of growth in trade with Arab countries, although from a very small base. So while Tunisia is geographically marginal to Europe it is highly dependent on Europe for its international trade.

Tunisia's imports are substantially above its exports: in 2002 imports were 13.5 billion TD and exports were 9.7 billion TD. This trade deficit is largely, but not totally, funded by tourism revenues, by remittances from Tunisian workers abroad (Rivlin 2001) and by foreign investment.

Table 5.2   Tunisia's main trading partners for exports and imports by value, 1999

EXPORTS			IMPORTS		
Country	Value million TD	Per cent of total exports	Country	Value million TD	Per cent of total imports
1 France	1835	26.3	1 France	2687	26.7
2 Italy	1575	22.6	2 Italy	1858	18.5
3 Germany	975	14.0	3 Germany	1121	11.1
4 Belgium	401	5.8	4 United States	434	4.3
5 Spain	376	5.4	5 Spain	406	4.0
6 Libya	287	4.1	6 Belgium	383	3.8
7 The Netherlands	211	3.0	7 Libya	281	2.8
8 India	179	2.6	8 Japan	251	2.5
9 UK	121	1.7	9 The Netherlands	218	2.2
10 Turkey	77	1.1	10 UK	212	2.1

Source: INS 2000

Tunisia's relationship with Europe for exports, imports and for tourism was a strong influencing factor in Tunisia becoming the first southern Mediterranean country to sign an association agreement with the European Union (see also Chapter 8 of this book). The association agreement entered into force on 1 March 1998 and Tunisia received US$250 million in loans from the World Bank to help fund the implementation plans. The purpose of the loans was to fund employment services and retraining, help the private sector improve the quality of its products and enable the government to cope with the loss of revenue from import duties (Rivlin 2001).

Foreign investment

With a large negative balance of trade and a government policy of greater integration with the global economy, Tunisia has focused attention on foreign investment. The government has a Minister for Foreign Investment and a Foreign Investment Promotion Agency. The policy of foreign direct investment in Tunisia is designed to:

increase exports from Tunisia
create jobs in Tunisia
transfer technology into Tunisia
support regional development in Tunisia.

Most sectors of the economy are open to foreign direct investment, with the exception of land which is not open to foreign investment because of concerns over national sovereignty.

In 2001 the level of foreign direct investment in Tunisia was approximately 886 million TD, with particular concentrations in the oil and gas, electricity generation and automobile component sectors. As with trade relations, foreign investment is mainly from western Europe, although there is a growth in investment by Arab countries, in particular Saudi Arabia and Kuwait. The largest Arab investment comes from the oil-rich Gulf states, for example in the building and the paper industries, and there is also some investment from Tunisia's North African neighbours.

Oil and Gas

The key to Tunisia's oil and gas sector lies in the geology south of the Atlas mountains. Tunisia's first oil field, El Borma, was discovered in 1964 in the southern region of the country near the border with Algeria. The El Borma field is an extension of the Ghadames basin in Algeria, a region of Triassic sandstones. Other oil discoveries have been made along the north and east sides of the Cretaceous carbonate platform in the centre-south of Tunisia, and offshore in the Gulf of Hammamet and in the Gulf of Gabes. Proven oil reserves in Tunisia are over 300 million barrels and production is about 85,000 barrels per day. Around 70 per cent of Tunisia's crude oil flows from the old but declining fields at El Borma, Ashtart and Sidi el Kilani.

Tunisia has an oil refinery at Bizerte that is operated by the Société Tunisienne des Industries du Raffinage (STIR). The refinery has a capacity of 34,000 barrels per day, less than half of Tunisia's daily oil production, so as a consequence Tunisia has to export oil and then import up to 50 per cent of its petroleum requirements.

The largest ever gas reserve found in Tunisia was the Miskar field in the Gulf of Gabes. It was discovered in 1974 with reserves of about 1 trillion cubic feet. The Miskar field is operated by the UK Company BG (a part of British Gas), which has also developed the Hasdrubal gas field: the Hasdrubal field will expand its gas production as the Miskar field reduces after 2009. The Miskar gas field provides 80 per cent of Tunisia's total gas requirements through the state utility company Société Tunisienne de l'Electricité et du Gaz (STEG), which operates a major electricity production and distribution plant located south of Sousse. Gas also comes from the El Borma oil field near the Algerian border, the El Franig and Baguel concessions in the south, and off-takes of gas from the TransMed pipeline that runs from Algeria through Tunisia to Italy.

In May 2003 BG announced that it had signed a Memorandum of Understanding with the Tunisian government in relation to the development of the US$250 million Barca Power Project. The project will use gas from the Miskar field to produce electricity and will be located adjacent to BG's Hannibal processing site about 21 km south of Sfax. In addition, BG plans to construct a Liquefied Petroleum Gas (LPG) plant at the Hannibal site to provide bottled gas to the Tunisian market.

The Importance of Tourism

The tourism sector is a significant contributor to the Tunisian economy. The measurable direct contribution of tourism to Tunisia's Gross Domestic Product is approximately 6 per cent per annum. In the year 2000 tourism revenues were about two billion TD in hard currency. In addition, there are indirect contributions to the economy through ancillary activities such as building, construction supplies, materials, transport, agriculture and craft industries. Approximately 300,000 people are in jobs directly or indirectly related to the tourism sector.

The attacks on the World Trade Centre of 11 September 2001 have had a strong negative effect on tourism in Tunisia, a negative effect that has been amplified by an attack on a synagogue in Djerba in 2002, the war in Iraq in 2003 and worldwide concern over the SARS virus during 2003. This section therefore examines the tourism sector before and after 2001.

Table 5.3 shows tourist indicator data for 1995 to 1999. Having invested over 400 million dinars in 1995 and 1996 to provide more hotels and other infrastructure there was an upward trend in the number of tourists and in the hard currency receipts from them in subsequent years. In the year 1999 the income from tourism was just short of two billion TD from nearly five million tourist entries. The pattern of growth continued into the new millennium: in the year 2000 there were over 5.3 million tourist entries to Tunisia, and in the first six months of 2001 there was an increase in tourism revenue of 13.2 per cent compared with the same period in 2000 (959.7 million TD compared to 848 million TD).

Table 5.3   Tourism indicators 1995–1999

Indicator	1995	1996	1997	1998	1999
Investment (million dinars)	449.1	417.0	353.3	307.0	323.0
Number of hotels	612	641	662	692	772
Tourism entries	4,119,800	3,884,600	4,263,100	4,717,500	4,831,700
Average stay (days)	5.7	6.2	6.5	6.1	6.9
Occupancy rate (per cent)	48.7	48.0	52.7	52.5	57.9
Hard currency receipts (million dinars)	1,322.9	1,411.0	1,565.3	1,712.8	1,950.0

Source: INS (2002)

In the year 2000 72 per cent of visitors came from western Europe (primarily France, Germany, Italy and the UK) and 27 per cent came from the North African countries of Libya, Algeria, Morocco and Mauritania. In mid-2001 there were expectations in Tunisia that the upward trend in visitor numbers would go on increasing at about 10 per cent per annum, with estimates of increases of tourists from France of 10-13 per cent, from Italy of 13 per cent and from Germany of 5-7 per cent, accompanied by expectations of substantial rises in tourism from eastern Europe, Russia and from Arab countries.

The events of 11 September 2001 brought a dramatic fall in the number of tourist entries to Tunisia. In January 2002 there were 34 per cent less tourists than in January 2001, and in the year 2002 as a whole tourist numbers dropped by over six per cent. This fall should be seen against the expected rise of 10 per cent. The main cause of the fall was fear by European tourists of visiting an Arab, Islamic country, rather than hostility by Tunisians to tourists, changes in visa requirements or reductions in hotel capacity. This fear is taking some time to diminish.

Table 5.4 compares the visitor/tourist entries to Tunisia for the months of January to October inclusive for 2001 and 2002. The information is organised by visitor region of origin. Visitor numbers were down over 20 per cent from Europe, with visitors from Germany declining by over one third. Visitor numbers from North America were substantially reduced, albeit from a very small initial base. The number of visitors from the Maghreb countries however increased by nearly 20 per cent, although it is clear from the data on bed-nights that these visitors were not tourists in the western sense but visitors to Tunisia for work or family reasons: for instance, the 1,109,539 Libyan visitors in 2002 only registered 259,096 bed-nights suggesting that they mainly stayed with relatives and friends, or with work placements.

Table 5.4   Visitors to Tunisia, January to October 2001 and 2002

Region/country of origin	Tourist entries January to October 2001	Tourist entries January to October 2002	Per cent change
Total European	3,388,815	2,671,615	-21.2
France	984,335	815,748	-17.1
Germany	868,557	558,700	-35.7
Italy	379,178	343,327	-9.4
UK	286,581	226,600	-20.9
Total Maghreb	1,440,905	1,723,849	+19.6
Algeria	519,616	575,911	+10.8
Libya	885,573	1,109,539	+25.3
Total North America	26,012	18,950	-27.2

Source: INS (2002)

Although Table 5.4 shows the decline in tourist visitors from Europe, there are still millions of tourists who visit Tunisia each year. While the majority visit for sun and sand, Tunisia is increasingly marketing itself as a location that offers a diversity of activities, including golf, tennis, thalassotherapy (sea water spas), desert tours and eco-tours. The Tunisian National Tourist Office (ONTT) promotes Tunisia as a country with a rich range of attractions alongside sun and sand: continuity and change in practice and presented as a consumer good.

Tunisia depends structurally on Europe, both in the sense of tourist visitors and their euros as a major source of national income, and in the sense of the tour operators that dominate the market. Companies such as Preussag (Germany), Alpitour (Italy) and MyTravel (UK) have strong leverage on the Tunisian tourist market by providing packages of air travel and hotel accommodation in which prices have been negotiated down to low levels. While there are direct flights from Tunis to the main European cities by scheduled airlines (Tunisair, British Airways, Air France, Lufthansa, etc), the majority of tourists arrive on the charter airlines of the tour operators.

Economic Development Plans

The Ninth Economic Development Plan

The Ninth Economic Development Plan ran for the five-year period 1997-2001. The overall goal of the Ninth Plan was the integration of the Tunisian economy in the international economic environment (see also the interview with President ben Ali in Chapter 14). Tunisia has joined the World Trade Organisation (WTO) and has an association agreement with the European Union. The Ninth Plan announced the liberalisation of the financial markets, including the goal of an open Tunisian dinar. The Plan had an economic growth rate target of 6 per cent per annum with associated investments in new technology, infrastructure and human resources.

Rural areas in Tunisia have had poorer infrastructure and facilities than urban areas for some decades. The Ninth Plan envisaged an increase from 76 per cent to 87 per cent in the proportion of the rural population with access to electricity, and an increase from 68 per cent to 78 per cent in the proportion of the rural population provided with safe drinking water. Agriculture attracted 16 per cent of all investment in the Ninth Plan, and table 5.5 shows the main investment categories. Investment in irrigation is clearly the largest category of investment. Water investments included the construction of new dams to create new water supplies and the designation and development of new irrigated perimeters for intensive agriculture.

Table 5.5   Investment in agriculture in the Eighth and Ninth Economic Development Plans

	Investment in million dinars (TD) at 1996 prices	
	Eighth Plan 1992-1996	Ninth Plan 1997-2001
Irrigation	1026.4	1771.1
Farm equipment	385.0	570.0
Water and soil conservation	206.3	230.9

The Ninth Plan also included action on the environment by reducing waste, reducing garbage and planting trees. A total of 1.6 per cent of national GDP was made available for ecology and environmental protection schemes. A mascot sprung up all over Tunisia as a reminder of concern for the environment. The mascot, a cross between a bat and a fox, is designed to encourage sensible waste disposal. In addition, all cities, towns and major villages have a *Boulevard de l'Environnement*, normally a tree-lined single or dual carriageway of smooth tarmac that is often in stark contrast to the roads leading up to and away from the boulevard.

The Tenth Economic Development Plan

The Tenth Economic Development Plan was approved by the Chamber of Deputies on 22 July 2002 and covers the five-year period 2002-2006. The goals of the Tenth Plan are to reinforce competitiveness, including competitiveness in the global economy, meet the challenge of employment and build a knowledge society/economy in Tunisia. This list of key goals is broader than the goals of the Ninth Plan but is commensurate with the Ninth Plan's goal of integrating the Tunisian economy within the international economic environment. The introduction to the Tenth Plan notes the difficult environmental and political context for the period of the plan, particularly citing the drought of 1998-2001 and the implications of 11 September 2001 for the Tunisian economy.

The Tenth Plan will invest around 30 billion TD in the Tunisian economy from public and private sources and has three main objectives.

1 To improve the financial strength of the economy as measured by fiscal indicators such as domestic savings and foreign balance of payments.
2 To achieve an annual growth rate of 5.5 per cent in the economy.
3 To create 380,000 new jobs and so reduce unemployment from 15 per cent in 2001 to 13.2 per cent by the end of the plan period. The creation of jobs in the knowledge economy will be highlighted, the proportion growing from 4.5 per cent to 9 per cent of all jobs.

Much of the Tenth Plan has a geographical focus. Access to power will be improved in rural areas by giving 95 per cent of the rural population access to electricity, up from 91.7 per cent in 2001. Access to safe drinking water will also be improved in rural areas with a plan to increase access to 80 per cent of the rural population. The basic infrastructure of roads, railways and industrial zones will be improved, including the building of 3800 km of new rural roads. Investments will be made in the health service and in social support, and there will be a focus on supporting the middle class who constitute 80 per cent of the working population.

Implications for the Sahel

Agriculture is an important part of the economy of the Sahel, so the problems of the agriculture sector have a direct impact on the region. In the Tenth Economic Development Plan the plan for agriculture growth is only 3.5 per cent per annum, compared to levels of 5-7 per cent per annum for industries and services, and 18 per cent per annum for telecommunications. The recovery from the drought years of 1998-2001 will help agriculture, but only from its relatively low current base in the Tunisian economy.

By contrast the Sahel region is a focus of much of Tunisia's tourism. In the area encompassing Sousse, Port el Kantaoui and Kairouan there are 108 hotels of which 11 are in the 5* category. These hotels provide several thousand jobs for Tunisian workers, and as long as there are no further shocks to the economic system there does appear to be the potential for further growth in western tourism to the Sahel and a return to the growth figures originally forecast in mid-2001.

Chapter 6

Population and Migration in Tunisia

Introduction

This chapter can be thought of as having two main parts. The first is concerned with describing and analysing trends in Tunisia's population. The second turns attention to international migration as it affects Tunisia. A theme that pervades the chapter, and is highlighted in its conclusion, is that its population and migration characteristics reinforce the idea that Tunisia lies 'at the margin of' or 'in between' Europe, Africa and the Middle East.

Population Size and Growth

Tunisia's population is relatively small. The total has increased from about 8.5 million in 1992 to about 9.5 million in 2001 (Table 6.1). This compares with populations in neighbouring Algeria and Libya of 30 million and 5.5 million respectively. It is striking at the same time that the rate of population growth has decreased significantly over the same period, from about a 2 per cent increase per year at the beginning of the 1990s to about a 1 per cent increase per year currently. This compares with the population explosion currently being experienced in sub-Saharan Africa to the south of Tunisia, and population decline to the north in Europe. And this observation reinforces once again the theme of Tunisia lying 'in between' Europe and Africa, in more ways simply than geographical location.

Table 6.1 Total population (thousands) in Tunisia (1992–2001) and percentage increase on previous year

Year	1992	1993	1994	1995	1996	1997	1998	1999	2000	2001
Population	8489	8657	8815	8957	9089	9214	9333	9455	9563	9673
Increase		2.0	1.8	1.6	1.5	1.4	1.3	1.3	1.1	1.1

Source: Institut National de la Statistique, 2002

The projection is for the continuation of this trend of decelerating population growth (Table 6.2). It is estimated by Tunisia's National Institute for Statistics that by 2029 the total population will stand at about 11.75 million, representing about a 2 per cent increase on the population estimate for five years before in 2024. By the 2030s it seems likely that Tunisia's population will stabilise at around 12 million, and probably start to decline slowly as is occurring across much of the industrialised world today. It will be interesting to see whether Tunisia adopts a similar response to a declining population as is currently being adopted in much of Europe and the USA, namely to encourage immigration.

Table 6.2   Projections for Tunisian population (thousands)

Year	2004	2009	2014	2019	2024	2029
Population	9947	10,388	10,814	11,210	11,538	11,763

Source: Institut National de la Statistique, 2002

Population, Age and Gender

The main reason Tunisia's population continues to grow, albeit at a slowing rate, and looks likely to continue to grow for the next thirty years or so, is that the majority of the population is young (Table 6.3). In 2001, 28.9 per cent of the population was under the age of 15. Rather frustratingly, the Tunisian National Institute for Statistics groups together in the single age category 15-59 the majority of the population (about 60 per cent), however there is no doubt that the majority of people in this category are concentrated in the younger years. Clearly the population is aging – in 1990 36.7 per cent of the population was under the age of 14 – but by no means at a rate sufficient to undermine population growth. As a corollary, there has been a gradual increase in the proportion of the population over the age of 60 – from 7.7 per cent in 1990 to 9.1 per cent in 2001. Life expectancy in Tunisia has increased from 70.6 years in 1991 to 72.9 years in 2001 (Institut National de la Statistique, 2002).

Table 6.3   The age structure of Tunisia's population (%) (1990–2001)

Age	1990	1991	1992	1993	1994	1995	1996	1997	1998	1999	2000	2001
0-4	12.4	12.1	11.7	11.4	11.1	10.6	10.2	9.8	9.4	9.0	8.6	8.3
5-14	24.3	24.2	24.0	23.9	23.8	23.4	23.0	22.6	22.3	21.9	21.3	20.6
15-59	55.6	55.9	56.3	56.6	56.9	57.6	58.2	58.9	59.4	60.1	61.0	62.0
60+	7.7	7.8	8.0	8.1	8.2	8.4	8.6	8.7	8.9	9.0	9.1	9.1

Source: Institut National de la Statistique, 2002

The sexes are very evenly represented in Tunisia's population. In 2001 males comprised 50.6 per cent of the population, which was exactly the same proportion as in 1990 (Institut National de la Statistique, 2002). Unfortunately, data that match age with gender are not currently available, however there is no reason not to assume that both males and females are overwhelmingly young in Tunisia. The age of females, of course, is particularly important in forecasting population growth, as is expanded below.

Population Distribution

In terms of population distribution and density there is a vast difference between the highly populated and urbanized north-east and the largely empty desert south. This is another respect in which the Tunisian Sahel occupies an intermediate position, with a regional population close to the national average of 65 per square kilometre.

Tunisia's population is becoming increasingly urbanized – a trend it shares with most countries in both northern and sub Saharan Africa (Zussman, 2000). In 1990 59.6 per cent of the population was defined by the National Institute for Statistics as living in urban areas. By 2002 this proportion had increased to 65 per cent. Net in-migration from rural areas and natural increase both contribute to the growth of the urban population. As an aside, already by the mid 1990s there were signs of labour shortages in rural Tunisia, to the extent that for several years the army was drafted to help with the olive harvest (Field, 1994).

The urban system is dominated by the capital city, Tunis (Table 6.4), with a population variously estimated from one to two million, depending on how many peripheral settlements are included. The second city (Sfax), the third (Sousse), and probably the fourth (Kairouan) are all in the Sahel zone, as are several medium-sized towns such as Monastir and Mahdia.

Table 6.4   The largest urban centres in Tunisia

Urban Centre	Population (000s)
Tunis agglomeration	1,600
Sfax	280
Sousse	180
Kairouan	130
Bizerte	120
Gabes	120
Gafsa	100

Source: Authors' estimates based on 1994 census data

The Demographic Transition

Trends in Tunisia's population indicate that it represents a fairly classic example of a country moving through the middle stage (stage two) of the 'demographic transition' (Box 6.1), characterised by a reduction in birth rates and steady death rates. Tunisia shares this character with most other Maghrebian countries (Sutton, 1999). In contrast, most sub-Saharan African countries are still in the first stage of the demographic transition – characterised by high birth rates and death rates, while many European countries have reached the end of the demographic transition, characterised by declining birth rates and growing death rates (hence population is declining).

Box 6.1 The demographic transition

There are three stages of the demographic transition. The first is characterised by high crude birth rates – typically above 40 per thousand, and high crude death rates – typically above 10 per thousand. This combination results in high rates of natural population growth, typically above 3 per cent per year. By stage two, crude birth rates have reduced to between 20 and 40 per thousand and crude death rates to between 5 and 10 per thousand. Population growth rates at this stage reduce, typically to between 1 and 3 per cent per year. At this stage population continues to grow, but at reduced rates. The final stage is characterised by low crude birth rates – at under 20 per thousand, and rebounding of crude death rates to similar levels to those in the First Stage. The outcome is population decline. Over the last twenty years, every Maghrebian country has moved from stage one to stage two (Sutton, 1999).

The relevant data for Tunisia are as follows. Fertility has fallen faster than almost anywhere in the world, from 6.5 in 1961 to 3.6 in 1991 and to 2.3 in 2001. In 1991 the crude birth rate was about 27 per thousand, whereas by 2001 it had reduced to 17 per thousand. Across the same period, death rates had remained fairly stable, at about 5.5 per thousand. As demonstrated above, the resulting population growth is about one per cent per year. As also alluded to above, the main reason that stage three will be delayed for Tunisia is a youthful (and therefore fertile) population, so that birth rates are not likely to reduce significantly yet.

Keith Sutton (1999) proposes three explanations for the Maghrebian countries having moved from stage one to stage two of the demographic transition over the past twenty years, each of which applies to Tunisia individually too. He depicts these as three 'revolutions' – urban, educational and social. Their impacts are inextricably linked. Put straightforwardly, increasing proportions of women are living in cities, where they have better access to education, which encourages awareness and often use of family planning. In addition, women in the urban workforce are likely to marry later, and pregnancy and childcare compete directly with paid work (Field, 1994). Sutton demonstrates the significance of this set of processes most tellingly using data from Morocco where, in 1987, the average rural woman had a family of nearly six children, while her urban counterpart had fewer than three children. Various data from Tunisia also reinforce the hypothesis. In 1994, for example, the fertility rate among illiterate women was 4.6, for women with primary education 3.0 and for those with secondary or higher education 2.4.

Sutton finally wonders whether there might be links between the emigration of Tunisians – the subject of the next section – and fertility levels among Tunisian women. First, as is demonstrated below, international migration to Europe is becoming increasingly feminised, such that for some women the opportunity to migrate may delay the decision to have children. Second, it is widely known that remittances – that is money sent home by migrants overseas – can change the behaviour of those receiving them. In particular remittances often increase consumerism, which again might compete directly with the costs of childbirth. Finally it is possible – though notoriously difficult actually to prove – that migrants can transmit home new codes of conduct (for example the idea of marrying later and having fewer children) that they have witnessed abroad (Levitt, 1998).

International Migration to and from Europe

This part of the chapter turns attention to international migration as it affects Tunisia. A number of reservations are worth posting straight away. First, the focus here is fairly exclusively on migration to and from Europe, as these are the flows for which relatively recent and reliable data are available. That is not to underestimate the significance of migration links with other parts of the world. In 1975, for example, the six Gulf oil exporters plus Libya had a workforce of about 1.6 million Maghrebians (Beaumont *et al.*, 1988; Findlay, 1994), and there are certainly significant numbers of Tunisians still working in the Gulf states today. Furthermore, it is estimated that between 1978-1981 about three quarters of a million Jews emigrated from South Asia and the Maghreb to Israel (Beaumont *et al.*, 1988), although there is no indication of the proportion that originated in Tunisia.

A second reservation is that there are similarly scant data on foreign nationals in Tunisia (other than tourists), and thus little can be said about immigration (for example by citizens of other Maghrebian countries or from sub-Saharan Africa) in Tunisia. Instead, the focus is exclusively on the international migration of Tunisians.

A third reservation is that even in Europe data on international migration are notoriously unreliable. They are often inaccurate, and usually hard to compare across different countries. The data discussed in this part of the chapter are mainly derived from Eurostat – the statistical office of the European Commission – that is widely agreed to be the best source on population and migration data in Europe. It is important, finally, also to add that the data used here record only legal migration. As is discussed later, it may well be that a significant proportion of Tunisian migration in Europe recently may not have been legal and has therefore not been recorded.

Tunisian Immigration in Europe

Table 6.5 shows the number of Tunisian citizens recorded across the 15 member states of the European Union. For most countries the data are for 2000, but for one or two they are more dated. Three countries stand out. France hosts by far and away the largest number of Tunisians in Europe – just over 150,000, which represents over 60 per cent of Tunisians in the EU. Some 55,000 Tunisians are resident in Italy, and a further 25,000 in Germany.

Table 6.5   Tunisian citizens in selected EU countries, 2000 or latest available data

Country	Tunisian Populations
Belgium	4159
Denmark	505
France	154,356
Germany	24,260
Greece	336
Italy	55,213
Netherlands	1312
Spain	590
Sweden	834
UK	2152
Total	243,892

Source: Eurostat, 2001

Put rather broadly, migration from Tunisia to these three countries illustrates the three main explanations that are usually provided to explain why migrants choose their destinations. One is historical links – in particular a colonial history. Most colonial powers provided for preferential migration from their ex-colonies after decolonisation. In addition, migrants from the ex-colonies are usually already familiar with the language and culture of the former colonial power. This is undoubtedly the main explanation for such a significant Tunisian presence in France. A second common way to explain migration patterns is proximity – in general most migrants do not move far. This may provide one of the reasons for the significant presence in Italy. The final main explanation is to do with work – most migrants go where they can work. Germany has traditionally recruited workers actively from the peripheries of Europe (most significantly from Turkey, but also from North Africa). In addition Germany has, until only recently, had a reputation for a strong and growing economy.

At the same time, it is worth noting that data on population stocks, such as those presented in Table 6.5, provide only a limited insight into migration patterns and processes. This is because they do not indicate when people arrived. The majority of Tunisians in France and Germany, for example, are not recent arrivals. Most arrived perhaps as long ago as 30 years, when preferential immigration rules in France and labour recruitment programmes in Germany attracted them. Furthermore, a good proportion of those recorded in the Table may never have migrated at all – they may well be the children or even grandchildren of earlier migrants, born in their host countries but maintaining their Tunisian citizenship. Until only recently, for example, even third and fourth generation migrants in Germany were not entitled to German citizenship.

A better picture of recent migration is provided by the data in Table 6.6, which are for officially recorded immigration in selected EU countries (not all countries provide these data to Eurostat). Again France, Germany and Italy stand out, but these data also reinforce the point made above that the majority of Tunisians in the EU have not arrived there recently. For example between 1990 and 1999, only 22,842 Tunisians legally entered France, representing only about 15 per cent of the total stocks recorded in France in 2000.

Table 6.6 Recorded immigration of Tunisian citizens to selected EU countries, 1990–1999

	1990	1991	1992	1993	1994	1995	1996	1997	1998	1999
Austria	:	:	:	:	:	:	171	164	142	159
Belgium	432	381	306	296	324	278	281	230	:	290
Finland	23	48	17	15	7	14	9	20	19	8
France	3847	4077	972	3447	2269	:	:	3281	4949	:
Germany	2664	2769	3096	2516	2312	2143	2002	1924	2261	2207
Italy	11540	5121	3160	:	1142	1146	5757	3348	:	:
Netherlands	297	290	226	189	162	173	146	200	170	128
Spain	28	26	27	21	15	18	10	33	34	47
Sweden	219	212	133	138	124	70	97	91	96	74

Source: Eurostat, 2001
: no data

Two further comments are worth making. Neither can be verified by data for the Tunisian context specifically, but both are common features of migration in Europe over the last two or three decades (Koser and Lutz, 1998). First, it is likely that an increasing proportion of Tunisian immigration in Europe over the last decade has been female. In general, young men dominate early migration waves. This would certainly have been true for those recruited to work, usually in industry, in Germany during the 1960s and early 1970s. After the 1973 'oil crisis', however, and more generally in reaction to an environment of recession, almost all formal labour recruitment programmes in Europe ended (some have been revived very recently, largely in response to population decline). Since the late 1970s, the main legal channel for entering Europe has been for family reunification – that is migrants have had the right to be joined in host countries by their immediate family. And in most cases this has been women (and children) following their husbands.

An associated point, again arising directly from the changing policy context, is that it is likely that illegal migration from Tunisia (and for that matter from most other countries outside the industrialised world) to Europe has increased significantly in recent years. By definition illegal migration cannot be recorded, but it is likely that the data in Table 6.6, which only record legal immigration, significantly underestimates total Tunisian immigration in Europe. Put rather simply, the reason that illegal migration has increased has been because for the past 20 years or so there has been virtually no legal way to enter Europe as a worker (Cornelius *et al.*, 1994).

Return Migration and Remittances

Data on return migration are even more patchy and unreliable than those on immigration. The main reason is that on the one hand many countries do not record who leaves, and on the other neither do they record returning nationals. In both cases this is because these migrations are not viewed as 'problematic' by governments. Tunisia does not publish data on returning nationals, and so the only indication available for return migration to Tunisia is from those selected countries in the EU that record emigration (Table 6.7). There is yet another reservation about even these data, however, which is that they demonstrate that Tunisians have left these countries, but do not guarantee that they necessarily went home. At least a proportion of those Tunisians recorded as emigrating from EU countries will not have gone home to Tunisia.

Table 6.7   Recorded emigration of Tunisian citizens from selected EU countries, 1990–1999

	1990	1991	1992	1993	1994	1995	1996	1997	1998	1999
Austria	:	:	:	:	:	:	112	110	88	60
Belgium	66	102	60	140	59	96	76	66	:	:
Denmark	19	30	17	12	10	17	9	7	18	:
Germany	1789	1789	1836	1924	2011	1949	1659	1565	1580	1249
Italy	97	105	116	:	108	205	215	246	:	:
Netherlands	66	67	62	60	39	71	67	50	65	33
Sweden	7	6	10	20	30	40	29	22	26	28

Source: Eurostat, 2001
: no data

What these rather unsatisfactory data indicate is that return migration has not taken place at a significant scale over the last decade or so, with the exception of from Germany. This is likely to cover the return of Tunisian workers brought to Germany on short-term labour contracts. But it is quite clear comparing Tables 6.6 and 6.7 that immigration in Europe by Tunisians has far outweighed their emigration from Europe over the past decade. Turning to return migration from elsewhere in the world, it is estimated that between 1984-1989 some 41,000 Tunisians returned from Libya – many of them expelled by Gaddafi (Cassarino, 2000).

A final observation worth making is that it is likely that Tunisians in Europe (and probably even more so those in the Gulf States) send home significant remittances. Once again there are no formal data with which to substantiate this assertion, but the Tunisians would be a very unusual migrant community if they did not remit money. Interviews by students in the field have confirmed that remittances are received in Tunisia, and it seems reasonable to surmise that extensions to houses, new cars and consumer items such as TVs may be a manifestation of remittances in Tunisia. Nevertheless, remittances remain notoriously difficult to measure, and their impacts equally difficult to prove (Box 6.2).

Box 6.2 Remittances

Remittances are notoriously difficult to measure, in part because a considerable proportion is sent home informally outside the banking system. Nevertheless a common estimate is that on a global scale migrant remittances amount to some $100 billion per year – making them the second most valuable item in world trade after oil.

A variety of factors determine whether migrants send home remittances and on what scale. These include characteristics of the migrants themselves – age, gender, education, employment and length of stay are all significant variables; as well as conditions in the host and home country such as a stable economy and reliable political system.

An unresolved debate concerns the impact of remittances. At an individual level they increase the spending power of friends and family members who receive them. At a regional level, this can have the effect of exacerbating inequalities between families and groups. The national-level impact largely depends on how remittances are spent. Often they are spent on imported luxury goods such as cars and televisions, thus having only a marginal benefit for the national economy.

Conclusions

In a number of ways this chapter has reinforced the theme that Tunisia (and the rest of the Maghreb) lies 'at the margin of' or 'in between' Europe, Africa and the Middle East in a wider sense simply than location. Demographic projections for gradual population growth, for example, contrast starkly with the population explosion in sub-Saharan Africa and population decline in Europe. This 'population gradient' across the Mediterranean is one reason why both legal and illegal migration has increased, and this chapter has demonstrated previous and recent migrations that link Tunisia with Europe and the Middle East. As expanded in Chapter 9, however, there are few migration links in either direction with the countries of sub-Saharan Africa. In its population and migration characteristics Tunisia is largely comparable with the other countries in North Africa, and this is one example where the concept of the Maghreb as a geographical entity makes sense.

Chapter 7

Women in Tunisia

Introduction

Chapter 4 alluded to the overriding theme for this chapter, which is that since Independence enormously significant strides have been taken towards trying to guarantee genuine equality for women in Tunisia. According to the United Nations Development Programme (UNDP): 'Tunisia stands out as the most progressive nation on women's issues in the Arab region' (UNDP, 2003). The first part of this chapter largely describes the status of women, covering their legal status, their participation in education, work and politics, and an overview of women's institutions. The second, shorter part of the chapter is more discursive and more critical.

The Code of Personal Status 1956

As the following sections demonstrate, immediately upon Independence Habib Bourguiba initiated a series of reforms that significantly improved the status of women in the years to come in politics and the economy. For example, the mandate to vote was extended to women and equal pay for men and women was made statutory.

Perhaps his most startling set of reforms, however, was targeted on the more personal sphere of women's right within the family. Many of these were embodied in the Code of Personal Status, which was promulgated on 13 August 1956 (Table 7.1). Among its most significant rulings – polygamy was abolished, judicial divorce was instituted, equal rights were guaranteed for women in divorce and it was made contingent on their own consent, a minimum age for the marriage of girls was established at 17 and widows were given the right to custody of their minor children. As we have seen in Chapter 6, family planning was also introduced on a wide scale by Bourguiba, thus reinforcing women's equality within the family.

It is worth reiterating the almost revolutionary nature of the Code of Personal Status and Bourguiba's other reforms given their context. First, they were introduced in a largely traditional Islamic society, and second long before comparable reforms were introduced elsewhere in the Islamic world. Take a moment to remember that female slavery in Mauritania was only officially banned in the 1980s, that female genital mutilation is still widely practised across the Horn of Africa, and that women are still stoned to death in countries like Yemen for infidelity. Remember also that women were given the right to vote in Tunisia before they were in Switzerland.

Table 7.1 Summary of the Code of Personal Status

Chapters	Focus
I	Marriage
II	Divorce
III	Entitlements after divorce or the death of a husband
IV	Food assistance
V	Protection
VI	Family relations
VII	Measures regarding adopted children
VIII	Measures regarding missing children
IX	Succession
X	Internment and freedom
XI	Inheritance
XII	Aid and assistance

Another indication of the farsightedness of the Code of Personal Status was that it remained largely unchanged until 1993, some 37 years later. This was when President Ben Ali introduced a series of amendments to the code (12 July 1993). These included most significantly: a legal guarantee of the rights of divorced mothers in the affairs of their children, the creation of a central fund to support divorced mothers and their children, and the creation of a magistrate's court to oversee the rights of women and children more generally.

Women and Education

Statistics on female participation in education in Tunisia make a fairly impressive impact. In 2001, for example, 97 per cent of girls aged six years old were enrolled in school (Institut National de la Statistique, 2002). Oddly, data for higher levels of education are significantly dated. Nevertheless, they tell us that in 1995 the proportion of girls in school education as a whole reached 46.8 per cent. At secondary level the proportion of girls had increased from 32.4 per cent in 1976 to 48.3 per cent in 1995, and at University level from 25.8 per cent in 1975 to 41.0 per cent in 1994 (Institut National de la Statistique, 2002). Given that Tunisia's population is very equally distributed across the sexes at all age levels (Chapter 6), the implication of these statistics is that females are only slightly underrepresented in education, in that they do not comprise fully fifty per cent of those at school.

Of course more research would be needed to understand the real meaning of these statistics. For example, although almost all six year old girls are enrolled in schools, what proportion actually attends and how regularly? And what are girls in secondary schools being taught – are they being channelled into traditionally 'female' subjects such as cooking or embroidery? And arguably just as important as statistics and research on females in education is that on males. Are boys being taught to respect the opposite sex? But on the whole these are subtle arguments that should not detract from the clear progress that has been made in female education in Tunisia.

Women and Work

Of course another way to gauge the significance of education is to see how it feeds through into subsequent employment. Table 7.2 shows female participation rates in various sectors of the economy. One sector that is not included in the table is the military, but it is interesting to note that on December 26 2002 the Ministry of Defence announced that mandatory military service would be imposed on Tunisian women in 2003.

Table 7.2   Proportion of women employed in various economic sectors (2001)

Sector	Proportion of Female Employees (%)
Textiles	76
Industry	43
Medical	43
Services	24
Legal	23
Public Services	21
Technology	21

Source: Institut National de la Statistique, 2002

Once again analysis is rather hampered by a lack of meaningful data. What can be said on the basis of Table 7.2 is that women dominate only one sector, namely textiles. This is unsurprising, and reflects trends in many other societies where textiles are traditionally viewed as 'women's work'. Clearly women are also well represented in the industrial and medical sectors. But here data limitations restrict interpretation. It is not clear precisely what type of industry is being referred to – presumably women dominate in light industry and men in heavy industry. And what do women do in these sectors? Are the majority of women in the medical sector nurses and ancillary staff, or are they also well represented at higher clinical and management levels? Similar sorts of questions arise for the other sectors that appear.

Furthermore, it is a shame that there are no published data on women's participation in the rural sector. The limited literature suggests an interesting pattern – that during the 1970s and early 1980s female participation in agriculture increased significantly as men emigrated, but that as opportunities to emigrate reduced and men started to return, the temporary 'feminization of agriculture' was reversed in the late 1980s (Ferchiou, 1998).

One possible reason why it might be difficult to quantify female participation in rural areas is that their work there may often be hidden, or ignored in traditional classification systems. Thus rather than ploughing fields or repairing cloches or tending animals, they may be collecting olives or preparing food. Women's work may be similarly 'invisible' in other contexts too. A good example would be 'home working', where women are employed on a piecemeal (and often exploitative) basis to work from home. This represents an example of the informal sector, which is by definition hard to quantify. Field observations have demonstrated to us a burgeoning informal economy, for example touting in tourist areas or selling produce on a small scale in urban areas. And this impression is reinforced in the limited literature (e.g. Berry-Chikhaoui, 1998). But again there are no data to indicate to what extent women are involved in this sector.

There is one final way that the data presented in Table 7.2 certainly do not represent a true picture of women's (or men's) work in Tunisia. That is because a significant proportion of people in Tunisia probably work in several sectors of the economy. This may reflect seasonal work patterns – for example working in rural areas during the harvest and in urban areas in winter. It may also reflect multi-employment – that is people with more than one job (salesman by day, security guard by night and so on). But it also may reflect the different activities that any single job entails. Thus, for example, the same person may harvest olives, transport them to the local village or town then proceed to sell them there.

Women and Politics

If the limited data provide a fairly optimistic impression of women's participation in education and the economy, the outlook is less rosy in politics. Starting at the political centre, women comprised just 6.7 per cent of Tunisia's Chamber of Deputies in 2001, albeit an improvement on only 1.1 per cent in 1966. Similarly, their membership of the Central Committee of the Rassemblement Constitutionnel Démocratique (RCD – the Constitutional Assembly or Parliament) stood at 11 per cent in 2001, compared with 3.1 per cent in 1957. At the local level women's participation is marginally better. In 2001 they comprised 16.5 per cent of the membership of Municipal Councils, compared with only 1.3 per cent in 1957. The message is clear: things are improving, but not quickly enough.

Another way to assess the participation of women in politics is to consider how many who are eligible to vote actually do. According to the National Institute of Statistics, in the 1994 General Election, 1.2 million women exercised their right to vote, as compared with less than 300,000 in the 1989 election. No data are available for female participation in the 1999 election.

It is hard, however, to calculate with any accuracy what proportion of women who could vote actually did so in 1994. We know that in 1994 Tunisia's population was about 8.8 million, of which just under 50 per cent were females, amounting to about 4.4 million females. We also know that in 1994 57 per cent of Tunisia's population (both sexes) fell between the ages of 15 and 59. Assuming that this age distribution was similar for males and females, there were about 2.5 million females between the ages of 15 and 59 in 1994. Of whom of course only those aged eighteen or above were eligible to vote. Even if as many as half a million females fell between the ages of 15 and 17, then only 1.2 million women of an population eligible to vote of about 2 million (or 60 per cent) actually did so.

Women's Associations and Institutions

If the participation of women in politics in Tunisia is less impressive that their participation in other sectors, it is noteworthy that a series of institutions have been established to promote and reinforce their rights. Four stand out. First, there is a cabinet level Secretary of State for Women and the Family, who heads the Ministry for Women and the Family, created on 13 August 1992 by President Ben Ali. According to its 'mission statement', the Ministry is concerned with promoting the rights of women in politics; proposing, funding and overseeing projects on the integration of women in the economy and guaranteeing the rights of women in the personal sphere of the family.

A second institution is L'Union Nationale de la Femme Tunisienne (UNFT – National Union of Tunisian Women), which was established by Bourguiba and has been strengthened under Ben Ali. The Union's principal activities are to combat discrimination against women in the workplace and promote women's involvement in decision-making. Two final institutions of note are L'Association des Femmes Tunisiennes pour la Recherche et la Developpement (AFTURD – The Tunisian Women's Association for Research and Development) and La Centre de Recherches, d'Etudes, de Documentations et d'Information sur la Femme (CREDIF – Centre for Research, Study, Documentation and Information on Women). These latter institutions are mainly concerned with studying and documenting women's rights.

It might be possible to criticise the way that these various institutions function. But on the other hand their very existence is testament to the fact that women's rights are taken seriously in Tunisia.

Critical Reflections

There are three more critical issues worth concluding this chapter with. First, although it has been stressed throughout this chapter that great advances have been made on behalf of and by women in Tunisia, the job is not yet complete. This chapter started with a laudatory UNDP quote. The same document, however, goes on to caution while 'near gender equality' has been achieved, 'women's position in Tunisia can still be improved' (UNDP, 2003). Of course the same could be said of the UK and any other country in the world. What is true is that progress has eluded certain sectors of Tunisia's female population more than others. For example, most Berber women still live in a very patriarchal society (Brett and Fentress, 1996). It is probably true that rural women are less affected by the types of reforms described in this chapter than urban women. And women in the informal sector (including home workers) do not enjoy the legal protection and representation of those working in the formal sector.

A second issue worth further reflection is the wider position of women in Islam. It is interesting that the Code of Personal Status does not contain explicit references to Islam. According to UNDP, the Tunisian government in recent years has tried to develop a new phase of Islamic thinking – *ijtihad* – distinct from that in all other Islamic countries. Critically, this brings Tunisian law into accordance with international human rights standards as opposed to Islamic law, and this distinction has underpinned the drive towards gender equality.

A third related issue concerns the value judgements that often influence the opinions of non-Muslims on the role and status of women in Islam. Even in secular and progressive Tunisia, a first time visitor might wonder to what extent women really are equal. In the medinas, for example, there are few women to be seen in public places, and where they are seen they are often dressed traditionally and largely covered. The point is that an outsider might view this as the subjugation of women, whereas a Muslim Tunisian might present this as a mark of respect for women. Ultimately it is not possible to divorce gender relations from their cultural context.

Chapter 8

Tunisia's Relations with the European Union: History in Five Phases

Alun Jones, Department of Geography, University College London

Introduction

The geo-strategic importance of the Mediterranean region to the European Union (EU) is indisputable. The Mediterranean region has always played a decisive role in the attainment of European peace and security. It is a geographical space in which several civilisations and religions have influenced and enriched each other, and is a crossroads for multiple cultural, human and economic exchanges (Xenakis and Chryssochoou 2001). From the beginnings of the European integration project in the mid-1950s the European Union has recognised the importance of the Mediterranean region and, largely through an economic agenda, has tried to ensure that political stability and peaceful relations prevail. This chapter will examine the ways in which European Union policy towards the Mediterranean region has evolved, focussing specifically upon Tunisia's economic and political involvement in this European policy process. Tunisia's relationship with the EU is best considered in five phases, starting in 1956 and continuing to the present day.

For convenience and ease of reading, the term European Union or EU has been used throughout this chapter as a uniform term in place of the European Economic Community (EEC) and the European Community (EC). However, the term EU in strict usage refers to the Europe of the Treaty of European Union (Maastricht). The term Community is used in a more general way to describe the EU polity.

Phase 1: Forging Ties with the Nascent EU 1956-1969

The establishment of the EU in 1957 with its ambitious economic integration programme prompted many states in the Mediterranean region, particularly those with close historical ties to Community countries, to seek a continuation of their privileged trade access arrangements. Such was the position of Tunisia after independence from France in 1956. For France, laden with colonial and post-colonial responsibilities especially in Africa, the EU offered a vehicle for the financial sharing of its historical obligations. For the EU, approaches from states like Tunisia for trade concessions increased the EU's political and economic standing in the international political economy. These three dimensions reflect the complexity of understanding EU relations with the Mediterranean region more generally.

The Treaty of Rome which set up the EU explicitly recognised the overseas commitments of France. A protocol annexed to the Treaty made specific provisions for France's former dependent territories and, as a result, Tunisia became one of the focal points for EU trade concessions. This held the prospect for closer formal links with the EU through an association agreement that enabled Tunisia to benefit from EU aid and technical assistance. Given the dependence of the Tunisian economy upon trade links with the EU (over 52 per cent of its imports and 51 per cent of its exports were with the EU throughout the 1960s) the Tunisian government was acutely aware of the importance of its relations with the EU and sensitive to any changes in this relationship.

The first danger signal came with EU efforts to set up a common agricultural policy (CAP) to protect the interests of European farmers. Central to the CAP were price support measures and import protection from foreign competition. The CAP had serious implications for a Tunisian economy in which the agricultural sector dominated in both employment and revenue terms. The introduction of the CAP in July 1966 not only delayed the progress of EU – Tunisia Association talks but also took much of the steam out of Tunisian hopes for major agricultural trade concessions. The EU, having invested so much in securing agreement among its member states for a Common Agricultural Policy, was particularly unwilling to jeopardise the CAP by admitting at lower tariff rates competitive farm produce from Tunisia and other Mediterranean countries outside of the EU. Tunisia, like many Mediterranean countries sought closer formal links with the EU throughout the 1960s, leading the Community to be confronted by recurrent pressures for major trade concessions.

In 1969 Tunisia successfully concluded an Association agreement with the EU for a five-year period with the aim of eliminating some of the obstacles to trade between both parties. The Association agreement was seen as a first stage in the long-term goal of creating a free trade area between Tunisia and the EU. There were however a number of criticisms levelled at the agreement. While the EU granted free access to the Community market for Tunisian industrial exports (except refined petroleum products and cork), agricultural trade was far more restricted. Here, trade concessions varied from 20-100 per cent depending on CAP market regimes. Fruit, vegetables and olive oil, the mainstays of the Tunisian agricultural sector, were given only limited concessions by the EU as a result of pressures from Italian producers fearful of competition within the Community. This prompted many to argue that the Association agreement was protectionist in nature, a fact confirmed by a specific clause within it which allowed the EU or one of its member states to take the necessary trade protection measures should serious disturbance occur in any sector of the European economy. Moreover, the Association agreement compelled Tunisia to grant the EU reciprocal access to its markets, further tying Tunisia's economic development to decisions made in Europe. Cumulatively, these criticisms begged the question whether the EU-Tunisia Association agreement was essentially neo-colonial in nature, locking the North African state into a new form of political and economic dependence.

Phase 2: Tunisia and the EU's Mediterranean Policy

By the early 1970s the European Community had signed trade/association agreements with all the countries bordering the Mediterranean with the exception of Libya, Syria and Albania. The agreement with Tunisia was thus part of a disparate patchwork spanning the entire region. As Shlaim and Yannopoulos (1976, 4) commented 'this disjointed incrementalism would have to give way to a more systematic and coherent approach ... which would take into account the problems and needs of the region as a whole'. By 1972 the idea of an overall 'global' approach to the Mediterranean region was gaining currency in the EU for several reasons (Lambert 1971). First, the proposed enlargement of the EU to include Denmark, Ireland and the United Kingdom meant that the existing agreements with states around the Mediterranean basin had to be re-signed by the new EU members. Secondly, many of the agreements, including that with Tunisia, would be expiring in 1974 and therefore up for renegotiation. Thirdly, the mosaic of agreements which had been concluded with Mediterranean states was signed at various stages of European economic development and union and in many respects reflected the emerging nature of the EU's institutions and policies. Fourthly, the EU was conscious of the geo-strategic significance of the Mediterranean region in the global confrontation between east and west, was aware of the region's importance in trade and investment terms, and was also a main supplier of labour, oil and minerals (Tsoukalis 1977).

The EU's Global Mediterranean Policy (GMP) was initiated in October 1972 by EU leaders and was to be pursued in three principal ways: free trade in industrial goods, the removal of restrictions on a substantial part of agricultural trade, and financial and technical cooperation. The problems encountered by the EU in formulating and implementing a GMP were particularly in evidence in Community negotiations with Tunisia. Agreeing concessions in the agricultural sector was the biggest stumbling block. Arguments over the level of tariff reductions for Tunisian agricultural exports, particularly olive oil and citrus fruits, hampered the progress of the negotiations for over three years. Secondary disputes, for example over Tunisian fishing access to Italian waters led to further delays. Equally problematic was the amount of financial and technical assistance the EU was prepared to grant Tunisia. For its part, the Tunisian government sought large-scale funding to finance ambitious development programmes. This presented major difficulties for the EU since it was also faced with a raft of similar claims from other states in the Mediterranean region, as well as pressures from its own member states seeking some compensatory redress for agreeing to the GMP.

As a result of these thorny issues the EU's Trade and Cooperation Agreement, as it was to be termed, with Tunisia was not signed until late 1976. The accord was hailed by the EU as a 'significant step in the relations between Tunisia and the EU ... which would contribute to [Tunisia's] economic and social development'. There was however no mistaking the disappointment felt by the Tunisian side regarding the agreement, despite the gloss put on the outcome by the Tunisian Minister for Foreign Affairs who commented in 1976 '... all agreements are the result of compromise, so it is not possible to cover all the aspirations of a country ... it is above all in the agricultural sector that the withdrawal of Europe has been the most marked' (European Commission 1976). However, there were serious anxieties over whether the EU would keep to the spirit of the agreement and respect the concessions that had been granted to Tunisia. As the Tunisian Foreign Minister explained '[We are concerned that the concessions granted could] suffer a slow erosion by unilateral regulations and protections ... which would risk emptying the agreement of a large part of its substance'. Analysis of the Trade and Cooperation agreement signed with Tunisia and similar agreements signed under the GMP with other Mediterranean states suggested that there were likely to be serious doubts over whether the EU could fulfil its obligations to Tunisia. EU agreements with both Spain and Israel included additional concessions for their agricultural exports to the EU, a fact likely to erode the Tunisian share of the EU market, particularly for citrus fruits. Also, the decision by the EU to implement specific calendar periods for Tunisian agricultural exports to the EU severely restricted the development of Tunisian export potential. For example, Tunisian tomatoes would benefit from a capped 60 per cent tariff concession by the EU only during the period November to April, and similarly potato exports only between January and March.

As regards the majority of Tunisian industrial exports, the EU had agreed that they could enter the Community market free of duties and quotas though the Community, and its member states reserved the right to impose import restrictions should difficulties arise in their own domestic markets. Concerning financial assistance the EU allocated 95 million € to Tunisia, though only 15 million € of this was in the form of grant aid. A number of projects were drawn up by the Tunisian authorities for EU assistance. These included rail improvements between Gafsa and Gabès, expansion of the port at Sfax, mechanisation of phosphate mining, development of irrigated farming, urban improvement studies and the promotion of tourism in the Tabarka area.

Cracks in the EU-Tunisia agreement began to appear within a year of its implementation. The EU, faced with mounting pressures in the operation of its CAP, rigidly enforced a system of quotas for several Tunisian agricultural exports. Important among these were apricots and table wines, products in chronic surplus within the EU. On the industrial front, the EU also began to introduce highly protectionist measures against a number of manufactured products of particular importance to the Tunisian economy. Chief among these were textiles. Here, the EU imposed import ceilings on a range of sensitive textile exports (including carpets) from Tunisia. Under considerable pressure from the EU, the Tunisian government was obliged to sign a voluntary export limitation agreement with the Community in 1979 which in the first instance would run for two years, and restrict Tunisian textile exports to specific agreed levels. With textiles comprising 32 per cent of Tunisia's exports to the EU in 1979 these European developments did not bode well for Tunisia's long-term economic growth. Other storm clouds were also gathering on the economic horizon, as the EU considered applications for full Community membership from Greece, Portugal and Spain, some of Tunisia's main competitors in the Mediterranean region.

Phase 3: Tunisia and the Mediterranean Enlargement of the EU

The accession of Greece to the EU in January 1981 and the positive signal given by the EU to Portugal and Spain for their membership sparked a multitude of discussions both within and outside the Community. Despite the EU's claim that enlargement would strengthen the role which the Community was destined to play in the world, it also appreciated that there would be serious consequences for states like Tunisia. The Tunisian economy was particularly vulnerable given the country's dependence upon the EU market. For example, in the agricultural sector 71 per cent of Tunisia's production was geared to the Community market. For some crops there was a staggering degree of dependence upon the EU as an export market. For example, in 1981 the EU took 83 per cent of Tunisia's citrus production, 87 per cent of its olive oil, 90 per cent of its fresh vegetables, and 100 per cent of its potatoes. The likely dynamic impact upon Spanish and Portuguese agricultural production as a result of EU membership was especially worrying for the Tunisian government, given that Spain, even before EU membership, provided over 40 per cent of EU citrus requirements. Tighter controls on Tunisia's agricultural exports therefore seemed inevitable. Between 1980 and 1984 the EU responded by subjecting Tunisian farm exports to specific quotas, ceilings and other reference limits as pressures over the CAP intensified.

Acutely aware of the trade implications of Community enlargement, the EU increased the amount of financial assistance available to Tunisia under a new financial protocol. During the period 1981-1986 139 million € was allocated to Tunisia for the funding of some 34 development projects. The majority of this money was allocated to rural development and agricultural diversification programmes since these were seen by the EU as important mechanisms by which the Tunisian farm sector could complement rather than compete with EU agricultural producers. In 1986 a third financial protocol was signed with the EU allocating 224 million € to Tunisia for the period up to 1991, and priority was again given to measures to promote rural development, reduce food imports and improve rural environments.

Phase 4: Tunisia and a Renovated EU Mediterranean Policy

By the end of the 1980s the EU became increasingly concerned about the security situation in the Mediterranean region, the rise of radical Islamist parties, and the pressing social, economic and demographic problems of states in North Africa. Within the EU, France and Spain called for a renewed Community emphasis on tackling these manifest problems. In 1989, the European Commission questioned whether enough had been done to assist the Mediterranean region, and mooted the suggestion of a long-term strategy for guiding EU-Mediterranean relations. In 1990 the EU agreed an expanded programme of support for the Mediterranean region. The 'renovated' Mediterranean policy, as it was termed, comprised six principal elements:

1   support for structural adjustment in the economies of the Mediterranean region;
2   encouragement of private investment by European companies in the region;
3   increasing member state and EU financial assistance to the area;

4 maintaining (and improving where possible) access for Mediterranean products to the EU market;
5 consultation with the Mediterranean states over the EU's drive for a single market in the Community; and
6 strengthening in a formal way the EU's political and economic dialogue with the Mediterranean region.

The significance attached by the EU to the renovated Mediterranean policy was reflected in a much expanded financial aid package for the period 1991-1996. Some 4405 million € were allocated to the Mediterranean, and of this amount 284 million € were to be distributed to Tunisia. The Tunisian government had drawn up an extensive list of development programmes in which EU aid would play a significant role. These included water and soil conservation/management, targeted rural development initiatives in the Sfax area, labour retraining and professional accreditation schemes, and a number of commissioned projects such as the redevelopment of the Gafsa basin. Tunisia also participated in broader EU assistance schemes known as the Mediterranean networks: Med-Urbs, Med-Campus, Med-Invest and Med-Media. These networks facilitated a number of exchanges and cooperation meetings between politicians, planners, business representatives and academics in the EU and Tunisia. Environmental protection, energy resource management and urban transportation schemes have been the priority areas. In this last context for example, local officials from Tunis met with their counterparts from Lisbon, Brussels, Valencia and Istanbul to examine ways of developing light railway systems within the Tunis conurbation.

Despite these welcomed developments in EU-Mediterranean cooperation, the EU was anxious to find ways of accelerating economic development in North Africa in order to prevent major social uprising, the spread of radical groups and terrorist attacks in mainland Europe (Marks 1996). Riots in Algeria in 1988, the hijacking of a French airliner in Algiers in 1994 and bombing campaigns in France in 1994-1995 starkly brought home to the EU the urgency of a new initiative on the Mediterranean region.

The signing of the Treaty of European Union (the Maastricht Treaty) in 1991 and the desire to establish an EU common foreign and security policy led to a reassessment of EU Mediterranean policy (Ludlow 1994). In 1994 EU leaders requested the European Commission to make proposals for the future evolution of Community policy towards the Mediterranean region. The Commission presented its ideas on this subject in late 1994 and spring 1995 with the notion of a 'Euro-Mediterranean partnership' occupying centre stage. The partnership was to be based on two over-riding objectives:

1 to underpin political reforms and secure human rights and freedom of expression; and
2 to support economic and political reform so as to boost economic growth rates and, hence, Mediterranean living standards and employment levels.

While EU states were in agreement over the importance of such a partnership, there was less consensus amongst them over the specific measures required to achieve it. For northern members of the EU, such as Germany and the UK, the most effective way to assist countries like Tunisia would be trade concessions, while for southern EU member states, such as Spain and France, the preferred policy would be financial aid. This political debate sharply revealed the difficulties of opening up the EU market for Mediterranean products and the reluctance of the EU's wealthier members to bear the costs of Mediterranean policy.

The Euro-Mediterranean partnership would, it was hoped, lead to a zone of peace and stability as a result of close political dialogue, respect for democracy, good governance and human rights (Jones 1997). In economic terms, the EU wished to promote free trade with Mediterranean countries, which would involve their economic modernisation and greater competitiveness. The EU envisaged offering assistance for such economic restructuring and adjustment, with a long-term goal of creating one of the largest free trading areas in the world. Additionally, the EU hoped that a wide range of issues could be dealt with on a co-operative basis such as illegal immigration, drug trafficking, prevention of terrorism, environmental protection, and energy exploitation and supply.

Phase 5: Tunisia and the Euro-Mediterranean Partnership

The Euro-Mediterranean partnership package was unveiled in Barcelona in November 1995 at a special meeting of ministers from all the member states of the European Union and the Mediterranean (except Albania, Libya and former Yugoslavia). The partnership has three central pillars: political and security concerns, economic and financial cooperation and assistance, and social and cultural issues (Gillespie 1997). The long-term objective of the partnership is to achieve a common area of peace and stability through political and security dialogue, social and cultural exchange, and economic liberalisation. In this latter case, the Euro-Mediterranean partners agreed to implement a free trade area by the year 2010. The EU committed 4685 million € from its budget to support the objectives set out at Barcelona.

To achieve the objectives of what is termed the Barcelona process the EU began a round of renegotiation of the bilateral agreements it had signed with Mediterranean states in the 1970s and 1980s. A new generation of Association agreements was to be concluded with each of the partners. Tunisia was the first Mediterranean state to sign a Euro-Mediterranean Agreement (EMA) in 1995 under the Barcelona process. This came into operation in March 1998. The EMA with Tunisia includes the following dimensions: statements about basic principles (democracy and human rights), provision for political dialogue, agreement on the free circulation of goods, economic cooperation, social and cultural exchanges, financial measures and institutional arrangements. Under the EMA the EU has allocated 428 million € to Tunisia to help achieve agreed economic reforms, such as privatisation efforts, the development of business centres and measures to improve the competitiveness of the Tunisian economy (Licari 1998).

Dialogue between the EU and Tunisia over the EMA takes place in a number of fora. Important among these are the *15+12* (15 EU states + 12 Mediterranean partners) meetings which have occurred with some degree of regularity since 1995. In 1996, for example, five meetings were held in the context of political and security dialogue, and plans were drawn up for six sectoral ministerial meetings (tourism, water management, industry, energy, fishing and information technology) at which policy harmonisation efforts and knowledge sharing were to be facilitated. There have been a number of criticisms levelled at the Barcelona process despite the undeniable fact that it represents a qualitative and quantitative shift from previous EU policies towards the Mediterranean region. Tunisia, like most of the Mediterranean signatories, lacks any better or more viable alternative to closer association with the EU. The EMA it has signed with the Community fails to recognise or deal with a number of key social, demographic and economic interests of vital geopolitical concern to Tunisia. First, the EU has not addressed the complexity of the migration problem, has not been prepared to consider the free movement of people given European security concerns, and has avoided tackling the pressing demographic problems that confront acutely the Tunisian social economy. Only from a European perspective of policing does the migration issue acquire any significance in the EMA. Secondly, while the EMA envisages the creation of a free trade area by 2010, numerous protectionist obstacles are still in force at the EU frontier, not least tariffs on certain significant Tunisian agricultural exports. Liberalising agricultural trade between Tunisia and the EU remains a long way off from political realisation. Thirdly, the EMA will have serious implications for the restructuring of the Tunisian economy as businesses respond to European competition. Many have argued that the level of EU financial assistance to the country remains woefully inadequate given these projected circumstances. As one observer has commented (Romeo 1998, 31):

there is no common long-term vision of what the role of the EU should be; European involution in the Maghreb region has been limited in scope, its foreign policy based on consensus, and the EMP [EMA] policy expresses partly the 'lowest common denominator' of the individual policies of the EU states. Worst of all, the current response could be detrimental to a 'safer Mediterranean' as intended.

Conclusion

For the EU, the Mediterranean region represents one of several areas of key political and geo-strategic interest, and therefore a delicate balance has to be maintained between its commitments to these various sites of important geographical interest (Pienning 1997). Since the instigation of the Barcelona process in 1995 the EU has been faced with a number of internal and external difficulties including growing concerns over the operation of the CAP and pressures for its reform, difficulties over budgetary commitments and future spending allocations, requests for improved access to EU markets, and its own ambitious programme for EU expansion to include not only countries in central and eastern Europe but also Cyprus and Malta in the Mediterranean region.

Tunisia's position appears to be precarious given the degree to which the Tunisian economy is anchored to political and economic policy processes in the EU. For some observers the EMA will aggravate the trade imbalance between Tunisia and the EU (Tovias 1997), threaten Tunisian industrial performance (Giammusso 1999) and further disadvantage the country's poorest social groups (Monar 1998).

Chapter 9

How 'African' is Tunisia?

Anthony O'Connor, Department of Geography, University College London

Introduction

Tunisia is clearly part of the African continent in physical terms, but how far is it 'African' in cultural, economic and political terms? Are there significant similarities and interactions across the Sahara, or are Tunisia's affinities with both Europe and the Middle East much stronger? The same questions can be asked of the countries with which it is certainly most closely linked, Algeria and Morocco to the west and Libya and Egypt to the east. So, the Tunisian case is central to the broader question of how separate northern or Arab Africa is from the remainder of the continent. Indeed, it may also shed light on the related question of whether the whole African continent or 'Sub-Saharan Africa' is the more meaningful entity in human terms.

The word 'Africa' is of Latin origin, and was first used specifically for the part of the continent nearest to Rome, i.e. the northern section of present-day Tunisia, with its sophisticated civilisation centred on Carthage. It was used a little later for a wider area extending into present-day Algeria and Libya, and more vaguely for land known to exist far beyond (Mudimbe, 1994); but its application to the tropical heart of the continent is much more recent.

Today, however, for most people in Europe and elsewhere, to think of 'Africa' is to think of either the whole of the continent or just the mainly tropical lands south of the Sahara. My enquiries among UK university students, and also their counterparts in four Tropical African countries, indicate a roughly equal balance between these two. Among 100 UK students a clear majority considered 'Africa' to mean the whole continent, yet an equally clear majority thought of Tunisia as part of the Arab world rather than Africa. Similar ambivalence has been found among students in Uganda, some there suggesting that separating off North Africa is denying Africa its millennia of 'civilisation', especially in Egypt.

This chapter will explore these issues under three main headings. First, it reviews English-language publications that have a bearing on the main question of how 'African' is Tunisia, although it is notable that no book in English or French deals primarily or explicitly with either the similarities or the relationships between Tunisia and the rest of Africa. Not even a book chapter or journal article on the topic has been found. Similarly, writings on present-day links between all North Africa or the Maghreb and the rest of Africa are scattered, fragmentary or dated, for example Chikh et al. (1980).

Second, the chapter considers ways in which Tunisia is similar to, or different from, most African countries, especially with respect to indicators of development or human welfare. Even a quick glance at the tables in such publications as UNDP's annual *Human Development Report* or the World Bank's annual *World Development Report* indicates a sharp contrast between Tunisia and most of Tropical Africa (see Tables 9.1 and 9.2). Tunisia often ranks alongside many Latin American countries and sometimes lies very close to the world average: most countries south of the Sahara do not.

Third, the extent of interactions between Tunisia and the countries south of the Sahara are investigated. Some forms of interaction, such as legal imports and exports or scheduled airline flights, are fully and reliably documented. Others, including the emigration of people, are much more poorly documented. Institutionally, Tunisia has links both eastwards and southwards as a member of the Arab League and the African Union, but no equivalent formal link across the Mediterranean to Europe (see the discussion in Chapter 8 of the Barcelona process); yet in many other respects there is much more intense interaction with Europe than in other directions.

Four final introductory points should be made. One is that this chapter is essentially concerned with the present rather than with the past, although the theme of continuity and change is taken up, with some attempt to discern recent trends rather than presenting an entirely static picture. A second is that the chapter deals with Tunisia as a whole, rather than focusing on its Sahel zone or exploring variations across Tunisia. For many topics considered here data for just the Tunisian Sahel would probably not differ greatly from data for the whole country. A third is that there is an emphasis on themes for which some degree of generalisation either for Sub-Saharan Africa or for Tropical Africa is possible, perhaps giving a false impression of homogeneity for these vast entities. And fourth, while most figures for Tunisia are thought to be fairly reliable, most for Sub-Saharan Africa are estimates liable to a large margin of error, and they should be taken only as indicating broad orders of magnitude. Availability of reliable data is indeed one striking contrast between Tunisia and most African countries.

Tunisia and Africa in English-Language Publications

Most encyclopaedia entries on Tunisia specify at the start that it is in North Africa, but then make no further reference to Africa, whereas they often have much to say about the significance of its Mediterranean location. Most entries on Africa start by considering the whole continent, but many are then followed by entries on topics such as African Art which are confined to Africa south of the Sahara. 'African' as an adjective frequently seems to be less comprehensive than 'Africa' as a noun.

Most United Nations publications concerned with worldwide patterns now use Sub-Saharan Africa as one of their units, while including Tunisia in another unit labelled 'Middle East and North Africa', or (as in the UNDP *Human Development Report*) one labelled 'Arab states'. The World Bank likewise uses Sub-Saharan Africa and Middle East/North Africa as groupings, for example in the annual *World Development Report*. Its annual publication *African Development Indicators* covers the whole continent, but the data are grouped into a set for the five countries of North Africa and a set for the 45 others.

Academic texts on Africa are roughly equally divided between those that consider the whole continent and those that confine attention to the area south of the Sahara (or even just Tropical Africa). This distinction is not clear-cut, however, for some (e.g. Reader 1997) include material on all parts of the continent while emphasising Sub-Saharan Africa, leaving Tunisia very largely ignored. An interesting case of this is *The Africans* (Mazrui 1986), even though its author is keen to stress the unity of Africa and even toys with the idea of including the Arabian peninsula within it. Mazrui has a particular interest in the role of Islam in Africa, and of all the writing on contemporary Africa it is that on Islam which most frequently considers together Tunisia and parts of Tropical Africa (e.g. Rosander and Westlund 1997). The most important academic work for us to note here is perhaps the 8-volume *Unesco General History of Africa* (Ki-Zerbo et al 1980-90), where the unity of Africa is a key theme of the Preface, which takes a continental approach throughout, and which has been published in English, French and Arabic.

Academic books written on Tunisia in English over the past twenty years have invariably devoted very little attention to the African dimension of Tunisia's location. Three examples will illustrate this.

Tunisia by Kenneth Perkins (1986) has as its sub-title 'Crossroads of the Islamic and European Worlds' in preference to any reference to Africa. Chapter 12 of this book is on Tunisia's foreign relations, but those with Sub-Saharan Africa are confined to two paragraphs on p 168, and are described as having 'virtually ended'.

Tunisia: the political economy of reform, edited by I. W. Zartman (1991), is a collection of thirteen essays, of which only the last, by Deeb and Laipson, makes any reference to Sub-Saharan Africa. Even this essay, on foreign policy, is concerned almost entirely with the Maghreb, the Middle East and the European Union, stressing that 'Tunisia must balance European ties and orientation with Arab roots and responsibilities' (p 239).

In *Modern Tunisia* by Andrew Borowiec (1998) Sub-Saharan Africa is mentioned only on pages 91-92 in a chapter on foreign relationships. He refers to Tunisians' awareness of 'the continuing turmoil in Sub-Saharan Africa', and notes that by 1997 'the best hope in Tunis was the strengthening of Maghreb unity, together with solidifying the Mediterranean option'.

The writing in English most sharply focussed on political relationships between North Africa and Tropical Africa is a book on The *Arab-African Connection* by LeVine and Luke (1979), although this is not specifically concerned with Tunisia. These authors stressed the change from much Afro-Arab solidarity as colonial rule ended to a situation in which high oil prices in the 1970s allowed the Arab world to interact mainly with richer nations and often on its own terms. 'The Arabs moved to a better part of town leaving their former friends behind' (p 97).

Similarities and Contrasts

Overview

Apart from its continental location, what characteristics does Tunisia share with the majority of African countries? When this question is asked in Britain some answers are clearly Eurocentric, adding up only to Tunisia not being one of the small group of rich countries with a population of largely European origin, with a majority at least nominally Christian and with disproportionate political and economic power in the world. Characteristics that are shared with countries ranging from Chile to China certainly cannot be regarded as distinctively African.

Whether in terms of geology or soils, temperatures or rainfall, natural vegetation or patterns of disease, Tunisia's natural environment, while similar to that of Algeria and Morocco, is vastly different from that of most African countries. It is, for example, not ravaged by malaria. There is nothing particularly African about the overall density or spatial distribution of its population: two-thirds of the population are urban dwellers, far more than in most African countries. Whatever we might mean by 'ethnicity' there is a clear ethnic distinction between the majority of Tunisians and most of the people who live south of the Sahara. It is unlikely, for example, that anyone of Tunisian ancestry in the United States would be categorized as 'Afro-American'.

One group of British students largely agreed that 'to think of Africa is to think of poverty', and this provided the starting point of a book on *Poverty in Africa* (O'Connor 1991). However an entirely valid criticism of that book is that it should have been titled Poverty in Tropical Africa. Few people anywhere would say that 'to think of Tunisia is to think of poverty', or of political conflict, of AIDS, or indeed of many of the negative images of Africa offered as alternatives to poverty. In an important political speech in 2001, the British Prime Minister Tony Blair stated that 'Africa is a scar on the conscience of the world': he repeated the phrase at the Earth Summit in Johannesburg in 2002. There is little doubt that he was not thinking of Tunisia.

In human terms, the most significant characteristic that Tunisia shares with most African countries is perhaps the legacy of a particular form and duration of colonial rule. As across much of the continent, French administration began in the 1880s; while independence was achieved in 1956, the same year as in Sudan, one year before Ghana and four years before most of francophone Africa. One result is a substantial element of duality in many aspects of Tunisian life that is also found across the continent from Mauritania and Senegal to Mozambique and Swaziland. A clear example of this is the co-existence of French and Arabic as the principal languages of the country. Others range from styles of dress to the physical structure of the cities.

Religion is a less appropriate example of such duality, since the Christian element is extremely small and Islam so clearly dominant. However, some would see Islam as itself a major link between Tunisia and Tropical Africa. Undoubtedly it is the basis of many cultural similarities between Tunisia and a dozen African countries from Senegal to Somalia, including two-thirds of both Nigeria and Sudan. It also means that there are cultural as well as environmental characteristics shared by the Tunisian Sahel and the West African Sahel. A more specific similarity between Tunisia and several countries in both western and eastern Africa is widespread adherence to a specific form of Islam known as Sufism – a form far removed from the Islamic fundamentalism so prominent in world affairs at the start of the 21st Century.

Of course, neither colonial rule ending half a century ago nor adherence to the Islamic faith is exclusively African, and it might be argued that Tunisia has just as much in common with Indonesia or Pakistan as with Burkina Faso or Tanzania. But at least the contrasts between Tunisia and most African countries are not as great with respect to religion and colonial legacy as for most of the characteristics now to be considered.

Development and welfare indicators

When the world is arbitrarily divided into 'developed' and 'developing' countries there is never any doubt that Tunisia, like all countries south of the Sahara, falls in the latter category. There is no ambivalence, as there might be for example, for Malta or Turkey. Tunisia is certainly in all senses of the word a 'developing' country: what is far less clear is whether that word is really appropriate for most countries in Tropical Africa at present. If our dividing line is between the top third on most development indicators and the bottom two-thirds, it must be drawn through the Mediterranean; but if it were to be between the top two-thirds and the bottom one-third, it would have to be drawn through the Sahara. And if, instead of insisting on a simple dichotomy, we think of the world in terms of rich, middle-income and poor countries, there is no doubt that again Tunisia and most Tropical African countries fall in different categories. The situation in respect of various indicators is shown in Table 9.1.

Table 9.1   Development and welfare indicators 2000–2001

	World	Tunisia	Sub-Saharan Africa
Gross national income per capita at purchasing power parity (US$)	7400	6500	1600
Life expectancy (years)	67	73	47
Infant mortality (per 1000)	56	21	107
Under 5 mortality (per 1000)	80	27	170
Adult literacy (%)	76	72	60
Human Development Index	0.72	0.74	0.47

Sources: African Development Bank (2002), UNDP (2003), World Bank (2002, 2003)

The most basic measure of prosperity is Gross National Income per head, which is now just over US$ 2000 for Tunisia, compared with under $500 for Sub-Saharan Africa and only US$300 for Tropical Africa (i.e. excluding South Africa). For Mali, Niger and Chad, the three Sub-Saharan countries nearest to Tunisia, and the heart of the West African Sahel, GNI per head is below US$200 (see Box 9.1). A more sophisticated, and arguably more realistic and useful measure, converts GNI to purchasing power parity to reflect different costs of living. In this case the Tropical Africa figure rises to US$1100, and that for Sub-Saharan Africa to US$1600; but that for Tunisia rises to $6500, comparable to both the figures for Latin America and the world average. Increased attention is now being given internationally to poverty as measured by the proportion of the population living on less than US$ 1 per day. This is officially estimated at between 30 per cent and 50 per cent in most Tropical African countries, but at less than 2 per cent in Tunisia.

Box 9.1 Economic profile: Tunisia and Niger

The contrasts between Tunisia and the nearest Tropical African country, the Niger Republic, are especially clear:

Tunisia	Data for 2000 or 2001	Niger
10	Population (million)	11
2100	Gross National Income per head (USD)	175
6400	GNI per head: purchasing power parity (USD)	890
+3.0	GNI per head annual % change 1988-2000	-1.5
10	% population living on under $2 per day	85
6600	Exports $m	260
9500	Imports $m	430
750	Foreign Direct Investment ($m)	15
190	TV sets per 1000 people	25
110	Telephone lines per 1000 people	2
15.3	Personal computers per 1000 people	0.4

Sources: UNDP (2003), World Bank (2002)

Since the 1990s increased attention has also been paid to composite measures of well-being which incorporate welfare indicators such as infant mortality and life expectancy. For instance, the United Nations Development Programme (UNDP) produces annually a Human Development Index on which Tunisia is ranked about 90, while most countries in Tropical Africa are in the range 140 to 175.

Whereas on most income indicators Tunisia stands rather below the world average, on most social welfare indicators it is above average, vastly different from most countries south of the Sahara. Current estimates of infant mortality are still above 100 per 1000 for most Tropical African countries, and the world average is just over 50. The Tunisian figure for 2001 in UN and World Bank sources is 21 per 1000, lower than the Latin American average and similar to the figure for Russia. The pattern is very similar for under-5 mortality, the key indicator used by the United Nations Children's Fund (UNICEF) (see Table 9.1).

Data in the UNDP *Human Development Report* for 2003 suggest that only 5 per cent of those born in Tunisia in 2000 are unlikely to live to the age of 40, contrasting sharply with 45 to 55 per cent in Tropical Africa. Life expectancy at birth in Tunisia is now reckoned to be 73 years, only a little lower than in much of

Europe, and higher than both the Latin American and the world average. The equivalent figure for Sub-Saharan Africa is only 47 years, and is currently falling due to the devastating impact of AIDS. Indeed, the incidence of HIV/AIDS is one of the most striking contrasts between Tunisia and most parts of Africa (and in this case especially southern Africa), for in Tunisia it appears to be among the world's lowest. Limited extent of cross-Saharan linkages may be relevant here.

On all education indicators Tunisia again compares well with the world average and is in a completely different league from most of Tropical Africa – even before quality of education is taken into account. Even for Sub-Saharan Africa as a whole, including South Africa, only about 25 per cent of secondary age children are in school, whereas the figure for Tunisia is around 70 per cent, which is similar to the world average. Figures for overall adult literacy are also similar to the world average, and far superior to those for most Tropical African countries, although there is a larger gender disparity than for the world as a whole, reflecting seriously deficient education for girls in the past.

Development and welfare trends

In relation to their frequently shared designation as 'developing countries', the most striking contrast between Tunisia and most Sub-Saharan African countries relates not to their present conditions but to trends over the past thirty years (Table 9.2). In 1960 life expectancy at birth in Tunisia was only 48 years, compared with the Sub-Saharan average of 40 years, and around 45 years in various individual countries such as Ghana and Kenya. The figure for Tunisia has since advanced by 25 years, while that for Sub-Saharan Africa has advanced by only 7 years. In 1960 under-5 mortality in Tunisia was over 240 per 1000, almost as high as the Sub-Saharan average, and higher than in countries such as Zambia, Tanzania, Uganda and Ghana. It had fallen to 80 per 1000 by 1989, and has since fallen steeply to 27 per 1000: meanwhile the Sub-Saharan average remains as high as 170 per 1000, contributing of course to the continued low life expectancy there.

During the 1970s per capita income rose in Tunisia over 4 per cent annually, compared with 1.5 per cent in Sub-Saharan Africa. In the 1980s the rate fell to 1 per cent, but Sub-Saharan Africa was then experiencing decline of 1.5 per cent a year. The contrast persisted through the 1990s – recovery to a 3 per cent annual rise in Tunisia, but near stagnation in Sub-Saharan Africa.

Another example of contrasting trends is provided by calorie consumption, which rose in Tunisia through the 1970s and 1980s from 2200 to 3000 calories per head per day, but which remained almost unchanged over Sub-Saharan Africa as a whole at around 2100 (according to UN estimates), and which fell substantially in various individual countries.

Table 9.2 Development and welfare trends

	World	Tunisia	Sub-Saharan Africa
Annual % change in per capita gross domestic product			
1975-2000	+1.2	+2.0	-0.9
1990-2000	+1.2	+3.0	-0.3
Life expectancy (years)			
1960	53	48	40
1975	61	58	46
1987	65	66	51
2000	67	73	47
Under 5 mortality (per 1000)			
1960	190	240	260
1987	110	80	185
2001	80	27	170

Sources: African Development Bank (2002), UNDP (2003), UNICEF (2002), World Bank (2002, 2003)

Demography

One aspect of Tunisia that is to some extent both cause and consequence of its economic and social development is the extent of recent change in its demographic structures and processes. Forty years ago Tunisia was very similar demographically to many Tropical African countries (Table 9.3). The crude birth rate was about 45 per thousand and the fertility level was about 6.5. With a death rate of about 20 per thousand, and some net emigration, population was then increasing at around 2.3 per cent per year. As a result of falling death rates, natural increase rose to a peak of about 2.6 per cent in the early 1970s, almost the same as for Africa as a whole, for Sub-Saharan Africa, or for many individual countries such as Ghana or Ethiopia.

By the year 2000, however, annual natural population increase in Tunisia had fallen to about 1.5 per cent, whereas over most of Tropical Africa it remained at 2.5 per cent to 3 per cent a year. The death rate in Tunisia had fallen to 6 per 1000; and fertility had fallen to 2.3 children per woman – compared with a world average of 2.8 and with a fertility level of 6 to 7 per woman in most Tropical African countries. The United Nations forecasts natural increase of only one per cent annually between 2000 and 2015, the same as for the world as a whole, compared with still 2.5 per cent over most of Tropical Africa. A very clear contrast is therefore now evident both in the current demographic situation and in recent and prospective trends.

Table 9.3 Demographic indicators

	World	Tunisia	Sub-Saharan Africa
Annual population increase (%)			
1975-2000	1.6	2.0	2.8
1990-2000	1.4	1.6	2.6
Fertility rate per woman			
1960	4.5	6.5	6.8
1990	3.1	3.6	6.2
2000	2.8	2.3	5.8
% population under 15 years			
2000	30	30	45

Sources: UNICEF (2002), UNDP (2003)

A closely related contrast can now be seen in the age structure of the population. In Tunisia in 1975 children under 15 made up 44 per cent of the total population, compared with 46 per cent in Sub-Saharan Africa and 48 per cent in Tropical Africa. By 2000 this proportion had fallen to 30 per cent, similar to the world average, it is expected to be only 23 per cent by around 2015, compared with 42 per cent in Sub-Saharan Africa and 44 per cent in Tropical Africa at that time.

Government effectiveness

Another set of phenomena that may be seen partly as a favourable condition for economic and social development, and partly as a consequence of it, are essentially political in nature. Perhaps in terms of the political system and civil liberties Tunisia has more in common with most of the Middle East than with either Europe or Sub-Saharan Africa; but the contrast with Sub-Saharan Africa is even greater with respect to the rule of law and government capability. In these cases there is a much clearer dividing line through the Sahara than through the Mediterranean.

The UNDP *Human Development Report* for 2002 presents subjective evaluations of countries throughout the world in respect of the rule of law on a scale of +2.5 to -2.5. Tunisia is awarded +0.8, compared with -1 to -2 for most Sub-Saharan African countries. In respect of government effectiveness, Tunisia is awarded +1.3, similar to France, Norway and New Zealand and well above Italy and Greece, compared with -1 to -1.5 for most of Sub-Saharan Africa. The very limited extent of the 'shadow' economy (i.e. exchanges taking place outside formally recognized channels) is one of the sharpest contrasts with most African countries.

An evaluation of the extent and severity of corruption is also provided in the 2002 *Human Development Report*. Again, Tunisia, with a score of +0.9 is better placed than Italy or Greece, and far better placed than Sub-Saharan African countries – most of whose scores range from -1 to -1.5. A similar picture is presented in the World Economic Forum's *African Competitiveness Report* for 2000/1. This report takes the whole continent as its framework, but contrasts across the Sahara are very evident throughout, and on many measures Tunisia has the highest ranking of all countries included in this report.

Trans-Saharan Interaction

External trade

The Trans-Saharan trade routes of the past included routes from Tunisia due south to present-day Niger and Mali. Much has been written on these in histories of Africa which take a continental approach (Davidson 1991, Fage 2002, Oliver 1999, Reader 1997). Today, however, these Trans-Saharan trade routes are of negligible importance and they have not been replaced by substantial sea or air movements of goods.

The share of Tropical African countries in the legal recorded external trade of Tunisia is tiny. Of the country's US$ 5900 million exports in 2000, over 80 per cent went to EU countries, 4 per cent went to Libya and 2 per cent went to the rest of North Africa (see also Table 4.2 in Chapter 4): the whole of Sub-Saharan Africa accounted for just under 1 per cent. Of the total imports worth US$ 8600 million, the EU accounted for over 70 per cent, USA for 5 per cent, the Middle East for 3 per cent, Libya for 3 per cent, and the rest of North Africa for 2 per cent. Sub-Saharan African countries accounted for well under 1 per cent.

There will be some unrecorded illegal trade, for example in narcotics, but there is no indication that this is extensive. Nor is there much evidence of the people trafficking between West Africa and Europe that has become an important element in the Moroccan second economy.

Migration

As shown in Chapter 6 of this book, there are large numbers of Tunisians in many European countries, and people of Tunisian nationality are also well represented in the USA and Canada as well as various Middle Eastern countries. But in no country south of the Sahara are there more than a handful. Thus the Arabs who play a significant role in the economy of Sierra Leone are from Lebanon and Syria, not from Tunisia.

Nor is there any substantial flow in the reverse direction, and certainly nothing to compare with the large numbers of Mauritanians, Senegalese and Malians in Morocco, or the many Nigerian migrant workers in Libya. It is remarkable that the 1984 census found only 38,000 foreign nationals in Tunisia, accounting for only 0.5 per cent of the total population. Of these, only 4,300 were not French, Italian or Arab. In the Tunisian Sahel, foreigners numbered only 3,800, and foreigners other than French, Italian or Arab only 600, or 0.03 per cent of the total population.

It is clear that these figures do not include tourists staying in the hotels of Sousse or Monastir; but while these tourists include many from Libya, and some from Saudi Arabia, it would be very hard to find any Ethiopian or even Nigerian among them.

Air routes

In 2001 there were scheduled international passenger flights from Tunis to 50 destinations. Only two of these, Dakar and Nouakchott, were in Sub-Saharan Africa (or only one if Mauritania is regarded as an extension of the North African Maghreb). There were frequent flights to Casablanca, Algiers, Tripoli and Cairo, and some to seven Middle East destinations; while 37 of the 50 destinations were in Europe (Table 9.4). The pattern had remained very stable throughout the 1990s, apart from the ending of scheduled flights to Bamako, Ougadougou and Accra in West Africa. In the case of some other North African cities there were more direct air links to Tropical Africa 30 years ago than there are today (in part reflecting a reduced need for fuelling stops on Europe – West Africa flights): but Tunis itself has never had many such links.

Table 9.4 **International air routes from Tunisia** (weekly outgoing scheduled flights from Tunis)

	Destinations			Total frequency		
	1969	**1985**	**2001**	**1969**	**1985**	**2001**
Europe	21	34	37	72	120	190
Middle East/SW Asia	2	10	7	2	17	21
North Africa	5	4	4	17	36	38
Africa south of the Sahara	1	2	2	1	4	2

Sources: ABC Airways Guide, OAG Flight Guide

Notes: Africa south of the Sahara refers to Monrovia (1969), Dakar, Khartoum (1985), Dakar, Nouakchott (2001). In intervening years there were brief periods with connections to Accra, Bamako, Kano, Ouagadougou

The extent of the focus on Europe is only reinforced by taking into consideration charter flights to Tunis and Monastir, nearly all of which come from, and return to, Europe. With respect to passenger numbers the reinforcement would probably be even greater, since most charter flights are filled more nearly to capacity than most scheduled services.

Diplomatic links

Diplomatically, as in most other ways, Tunisia looks northwards and eastwards rather than southwards. In 2002 57 countries had embassies in Tunis (Table 9.5), of which Sub-Saharan Africa accounted for only eight (Congo D R, Cote d'Ivoire, Djibouti, Mauritania, Senegal, Somalia, South Africa and Sudan). Europe accounted for 22, the Middle East for 11, and North Africa for four. Tunisian representation abroad shows a very similar pattern, including no representation in the nearest Tropical African countries, Niger and Chad: but it does maintain an embassy in Mali.

Table 9.5 Countries with embassies in Tunisia (2002)

Europe	22
SW Asia	11
Other Asia	6
Americas	6
North Africa	4
Africa south of the Sahara	8

Sources: Europa (2003)

Notes: The sub-Saharan African countries are Congo DR, Cote d'Ivoire, Djibouti, Mauritania, Senegal, Somalia, South Africa, Sudan. Of these, Tunisia does not have embassies in Djibouti or Somalia, but does in Cameroon and Mali

The number of Sub-Saharan African embassies is notably higher in Algeria, Morocco and Libya (19, 18 and 16 respectively) out of totals not a great deal larger than in Tunisia.

Institutional and Cultural Links

The international organisation of most importance for Tunisia since independence has been the Arab League, in which it played an especially active role in the 1980s before the headquarters moved from Tunis to Cairo. This currently has 22 members, including Mauritania, Sudan, Djibouti and Somalia; and it serves as an umbrella for various specialist bodies of which two, the Arab States Broadcasting Union and the Arab League Educational, Cultural and Scientific Organization, are based in Tunis. However, Tunisia is also a member of the Organization of the Islamic Conference, whose 57 members include 20 in Sub-Saharan Africa, and which assembles every three years to promote cultural solidarity. In 2001 it also joined the *Communanté des Etats Sahelo-Sahariens*, whose 18 members span the Sahara, but this is a much weaker organisation with very limited funds.

For four decades Tunisia combined membership of the Arab League with membership of the Organization of African Unity (OAU), and in 1994-5 Ben Ali served as OAU president: but by this time the OAU had become an extremely ineffective body, close to collapse. For many Tunisians, the 52 member *Confédération Africaine de Football* provided and continues to provide, a far more significant link with the countries south of the Sahara, especially countries such as Cameroon and Nigeria. Tunisia is due to host the final of the African Cup of Nations competition in 2004.

Foreign policy links

In terms of foreign policy towards Sub-Saharan Africa, Tunisia differs considerably from all of its North African neighbours, partly because it alone has no boundary with any Sub-Saharan country. Both Morocco and Algeria are very much concerned with the Western Sahara situation, albeit from opposite perspectives and have consequently much interest in relationships with Mauritania. Tunisia takes a much more neutral stance on the Western Sahara issue. The real contrast, however, is with Libya. Not only has Libya invaded and occupied part of Chad, but also its head of state, Colonel Qaddafi, has provided much financial as well as political support to leaders such as Idi Amin of Uganda, and has made clear his aspiration to leadership at an African continental scale. His performance at the 2002 meeting in South Africa to replace the Organisation of African Unity with the new African Union was a clear example of this. President ben Ali does not of course have the same resources from oil revenues to attempt anything similar, but while he is positive in principle to the development of the African Union his priorities clearly lie elsewhere. Almost fifty years ago Habib Bourguiba played a major role in Ghana's independence celebrations: 'He saw his own country as a relay station between Mediterranean civilisation and the emergent regimes to the south' (Chazan et al 1999: 377). 'Tunisia was the first of the Maghreb states to pursue an active policy in Sub Saharan Africa', but it was 'subsequently overshadowed by the initiatives of Libya, Algeria and Morocco' (ibid). Today it seems even more true for Tunisia than for its neighbours that 'North Africa is disconnecting itself from the rest of the continent' (Mbembe 2002: 68).

Above and Below the National Scale

This chapter has been concerned almost entirely with Tunisia as a national entity. I have been concerned with how far the country as a whole looks southwards as against eastwards or northwards to other world regions. Tunisia's greatest similarities, however, and its closest links, are with its North African neighbours; while in many respects there is great diversity within Tunisia itself. Bringing these two points together, there is no doubt that people in the parts of the Tunisian Sahel closest to the Algerian border have the closest affinity with that country.

On many of the indicators in this chapter Tunisia stands close not only to the world average, but also to the average for North Africa. To varying degrees, therefore, Morocco, Algeria, Libya and Egypt also stand apart from the rest of Africa; but each of these countries has at least one closer link. Morocco is an important transit country for West African migrants heading towards Europe. Morocco and Algeria are both deeply involved in political strife over Western Sahara. Libya's leader, Colonel Qaddafi, has both past involvements and future ambitions south of the Sahara. And the course of the River Nile ensures that Egypt has vital interests in Sudan, Uganda and Ethiopia.

This is not the place for a long discussion of regional diversity within Tunisia, but if a position close to the world average is a feature of the country as a whole it is also a feature of the Tunisian Sahel – to a greater extent than either Tunis city or the poorest rural areas of the country. No evidence has been found to suggest that any of the points made in this chapter do not apply to Sousse and its hinterland. Even when a topic relates specifically to Tunis, i.e. in the case of embassies or air routes, this is because the majority of the Sahel's international links operate through Tunis. Air routes direct to and from the Tunisian Sahel do exist in respect of charter flights, but it will be no surprise to learn that none of these is to or from Tropical Africa.

Conclusions

One constant in an ever-changing world is that Tunisia is fixed in the African continent. However, in human terms, Tunisia at the start of the 21st century seems much more strongly oriented both northwards to Europe and eastwards to the Middle East than southwards to Sub-Saharan Africa. Furthermore, most changes over the past three decades have increased the differences between Tunisia and most African countries – in most cases to Tunisia's great advantage. In development and welfare terms Tunisia has risen to a position close to the global average, while Sub-Saharan Africa has increasingly stood out as the world's most impoverished region.

Tunisia's links southwards seem even weaker than those of its neighbours, Libya, Algeria and Morocco, perhaps partly because among the four it is most clearly a country lying *north* of the Sahara, rather than being itself largely Saharan. At present there seems very little to attract Tunisians towards Sub–Saharan Africa, economically, culturally or politically (with the possible exception of Islamic solidarity in respect of parts of Tropical Africa): but it will be interesting to see whether the transformation of the OAU into the African Union brings any change.

The final conclusion must be that although there is much ambivalence in the published literature on Tunisia and Africa, and although Tunisia very nearly *was* 'Africa' from the perspective of Rome 2000 years ago, it has few similarities with, and few links with, most of Africa today. We can (cautiously) generalize on many topics and trends for Sub-Saharan Africa, and on more for Tropical Africa (distinct from both North and South Africa): but it is hard to find any topic for which we can make valid generalizations for the whole African continent, including Tunisia, unless they apply also far beyond, for instance to all poor and middle-income countries.

PART TWO
CONTINUITY AND CHANGE IN THE SAHEL

Chapter 10

Thysdrus and Ancient Rome

Introduction

The quiet town of El Djem was in Roman times the major settlement of Thysdrus. The town grew in the first century AD to reach a peak in the second and third centuries AD as one of the finest cities in the Roman world, and then declined after the third century AD. The wealth of Thysdrus came from a combination of two factors: its geographical location and the cultivation of olive trees.

Roman Influence

Invasion

Thysdrus first came to prominence as an important crossroads of north-south and east-west land routes. Figure 10.1 shows the ancient road network of the Thysdrus region, clearly illustrating the central role of Thysdrus in connecting the coastal towns such as Gummi (now Mahdia) and Taparura (now Sfax) with the Roman towns of the interior such as Sufetula (now Sbeitla) and Masclianae (now Hajeb el Aïoun).

When Julius Caesar landed at Ruspina (now Monastir) in the year 46 BC the people of Thysdrus offered 300,000 bushels of wheat as a peace offering (Slim 1996). Caesar rejected this peace offer and attacked and conquered the interior of what was then termed *Africa Nova* and *Africa Vetus*. The term *Ifriqya* was also used, subsequently changed to *Africa*. Under Roman rule Thysdrus was granted the status of a free city because it supported Rome actively during the Third Punic (or Carthaginian) War (Rogerson 1998). During the Roman occupation there was some integration of Romans with the local population and there is some evidence in Thysdrus of Romans worshipping Baal, a god of pre-Roman Africa.

Revolution of 238 AD

During the second century AD an African bourgeoisie developed in Thysdrus based on the agricultural wealth of the region, most notably in wheat cultivation and olive oil production. At that time Rome had a dedicated African grain fleet – the *Classis Commodiana Africana* – to ship wheat to Rome quickly from the granary of Africa Nova. Travelling at 15 knots the ships of the fleet could travel from the coastal town of Ruspina to the Italian coast near Rome in as little as 30 hours, clearly indicating the nearness of Tunisia to the heart of Europe in Roman times.

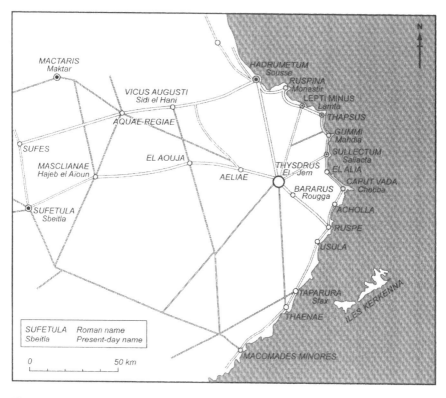

Figure 10.1   Ancient road network of the Thysdrus region
Source: adapted from Slim (1996)

However, resentment grew in Thysdrus at the high levels of taxation imposed by the Roman emperors and particularly by the Emperor Maximinus, originally a simple Thracian soldier. Local tax collectors in Thysdrus amplified the demands from Rome and were responsible for confiscating financial and other assets from the citizens and for stripping metal from statues as payment of taxes. One night in January 238 AD a group of Thysdrus citizens attacked the chief tax collector and stabbed him to death. They then went to Gordian, the Roman proconsul and a venerable Roman senator based in Thysdrus, robed him in a purple cloak and proclaimed him emperor.

Gordian I, and his co-emperor son Gordian II, briefly ruled the whole Roman empire in 238 AD from the governor's palace in Carthage (now the location of Tunisia's presidential palace), but did not secure the support of Capellianus the legate of the III Augusta Legion. Capellianus marched on Carthage and defeated a small army led by Gordian II, then carried out an extensive purge of the region of Africa Nova, plundering temples, seizing money, killing the gentry who had supported Gordian I and burning villages and fields (Rogerson 1998). Thysdrus came off particularly badly as the origin of the uprising against the Emperor

Maximinus, and Gordian I worst of all as he committed suicide on hearing of the death of his son Gordian II at the hands of Capellianus.

Amphitheatres

Today the most visible sign of Roman rule over the Thysdrus region is the amphitheatre in the centre of the town of El Djem. The amphitheatre, which is shown in figure 10.2, is the largest and most important ruin from the times of ancient Rome in Africa, and is in fact the third amphitheatre to be built in Thysdrus.

The first amphitheatre in Thysdrus was carved out of a hollow in the ground on a site immediately south of what is now the El Djem railway station. Rudimentary seating was carved out of the local tufa base rock. This first amphitheatre was filled in and a new, second amphitheatre was built on the same site. The new amphitheatre was an oval shape measururing 60m x 40m and seating was constructed out of brick and clay for 7000 spectators. The second amphitheatre is visible today as an elliptical mound with a flat central arena and some evidence of the structured layout of the original seating. It looks ripe for renovation and re-creation.

The third amphitheatre (see Figure 10.2) was built 500m north of the site of the first two on flat land, and is altogether a more ambitious project. The third amphitheatre was constructed during the first decades of the third century AD when Thysdrus was at the peak of its power. It was modelled on the Colosseum in Rome and is about three quarters the size of the Colosseum. The Thysdrus amphitheatre measures 148m x 122m in plan and is 36m high. The amphitheatre has three major levels of arches in its outer ring, and has raked seating for 30,000 spectators in four levels in the interior.

Figure 10.2 The El Djem amphitheatre today

Originally the amphitheatre also had a partial canopy to provide shade from the intense sun of the summer months, somewhat like a modern football stadium. The arena in the centre has entrances at each end and rectangular openings in the floor through which wooden platforms could be raised from below. Cells below the arena were used for storage, for keeping animals before their appearance in the arena and for drawing water from wells built below the amphitheatre. Figure 10.3 is a sketch of the likely appearance of the third amphitheatre at its largest and most active time.

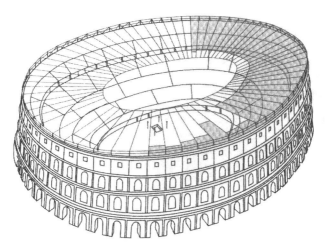

Figure 10.3   Sketch of the amphitheatre at its height
Source: adapted from Slim (1996)

The amphitheatre was used on occasions for gladiatorial and wrestling contests, but its more general use was as a place of entertainment. Its capacity of 30,000 showed its regional significance because the town of Thysdrus never housed that many people itself. The audience in the amphitheatre gathered to watch reconstructions of famous battles, presentations of wild animals, plays and marches, as well as sacrifices of prisoners and gladiatorial contests. The Roman mosaics from Thysdrus in the present-day museums in El Djem and Sousse (see Figure 10.4) illustrate richly the range and diversity of animals shown proudly in the amphitheatre. A common theme of the mosaics is fertility, bounty and fruitfulness, clearly showing the emphasis in the amphitheatre of reminders of the good life.

Wheat and Olives

During 1st century AD the Thysdrus area was successful in producing wheat, much of its shipped to Rome using, as noted above, the *Classis Commodiana Africana*. However, wheat productivity was relatively low because of the poor physical environment: irregular and low rainfall, intense summer heat, heavy clay soils and

Figure 10.4   A mosaic fishing scene in the Sousse museum

saline aquifers. An opportunity came in the 2nd century AD to shift away from wheat because of a reduction in wine and olive oil production in the northern Mediterranean regions. These reductions were caused by social and economic crises. In place of wheat fields olive trees were planted extensively in the Thysdrus region during the 2nd century AD, an agricultural revolution from wheat to olive oil that set Thysdrus on the road to become the oil capital of the Roman empire.

Olive oil was used throughout the Roman empire for cooking, as fat in the diet, as lighting fuel for lamps and as the basis for perfumes. Thysdrus became a thriving economic and social centre with extensive trade because of the wealth of olive oil and the town's geographical location on north-south and east-west trade routes. Along with trade with Rome, there is also evidence in Thysdrus of trade with the eastern Mediterranean and with Arabia. Thysdrus expanded its other small scale industry and trade, in particular metal working, plaster working, stone sculpture, pottery and jewellery.

The shift from wheat to olives made a profound and long lasting impact on the landscape of the Sahel region of Tunisia. The legacy of the olive tree planting can be seen now as regularly-spaced olive trees are lined up for mile after mile across the plains and rolling hills of the Sahel, and provide a clear statement of continuity in the landscape. Figure 10.5 shows the impressive and extensive impact the olive tree makes on the landscape of the Sahel.

Figure 10.5   Lines of olive trees

Decline and Resurgence

As the Roman empire declined so did the role and influence of Thysdrus. The Romans moved their African power base to Sufetula (now Sbeitla) 130 km to the west in the Dorsale mountains. Thysdrus declined to a small village, except for the imposing third amphitheatre. By the 6th century AD the amphitheatre had been turned into a fortress and is traditionally claimed (along with several other places in North Africa) to be the site of the last stand of the Berber queen Al Kahina against the Arab invasion. The new focus on the religious city of Kairouan inland and the port city of Sousse on the coast shifted the centre of power away from Thysdrus and the amphitheatre became a source of stone for new buildings, most notably the grand mosque in Kairouan. The amphitheatre remained largely intact until 1695 when artillery shells destroyed a major section of the northern wall during a battle.

During the 20th century the amphitheatre was restored to a state much better than the Colosseum in Rome. It is now run by the Tunisian national heritage agency and has been recognised by the United Nations Educational, Scientific and Cultural Organisation (UNESCO) as a World Heritage Site. It is a major tourist attraction, with plays, music and opera all staged in the amphitheatre several times a year, a direct reflection of its original function as a place of entertainment. The amphitheatre is often unknowingly seen by a wide audience as a location for films, television shows and television advertisements.

Box 10.1 Music in the Roman amphitheatre

In the summer of 2002 the 17th Symphonic Festival was held in the Roman amphitheatre in El Djem. The concert programme illustrates a combination of national and international links in music, and reflects the original use of the amphitheatre as a major place of entertainment.

Saturday 20 July 2002	Music and Solidarity, a concert by the Orchestra of Vienna Opera Ball in aid of the 26-26 National Solidarity Fund
Wednesday 24 July 2002	Tunisian Symphony Orchestra
Friday 26 July 2002	String Orchestra of the National Conservatory of Lyons
Saturday 27 July 2002	Musica nel Mondo, a recital of Italian opera arias
Saturday 3 August 2002	Rossini's The Barber of Seville performed by the Bolshoi Theatre from Russia
Sunday 10 August 2002	State Orchestra of Kazan

Conclusion

The imposing and dramatic amphitheatre in El Djem is a clear statement of continuity in Tunisia. It is a testament to the wealth of central Tunisia some 1800 years ago and is a reminder of the importance of olive trees and the theme of continuity in the landscape. Tunisia is now sensibly exploiting the Roman presence in Thysdrus for its cultural and historical benefits, a change that fits the needs of the 21st century.

Chapter 11

Agriculture and Fishing

Introduction

Agriculture in Tunisia is still mainly characterised by traditional products - olives, olive oil, wheat and other grain crops, dates, citrus fruits and dairy products. Innovative crops, such as red peppers, tomatoes and courgettes, are grown often for the tourist sector and for export. The thread of farming continuity, combined with rainfall shortages in recent years and competition from other Mediterranean countries, has produced an agricultural sector that is performing poorly, and in fact reduced its contribution to Tunisia's Gross Domestic Product in 2001 (see chapter 5). This chapter reviews the elements of continuity and change in Tunisian agriculture, including investments in water for agriculture, and then goes on to review the Tunisian fishing industry, an industry that has expanded to a current plateau in annual production.

Agriculture

Continuity and the olive

Agriculture in the Sahel clearly illustrates the two themes of continuity and change. Continuity in the Sahel is represented by olive tree cultivation, while change has been seen in agriculture since the start of the French protectorate when new land ownership arrangements were introduced.

Olive tree cultivation was introduced by the Romans in the 2nd century AD in the agricultural revolution from wheat to oil (see chapter 10). Olive trees are resistant to drought, easy to maintain and have become constant features in the Sahel landscape. They are at their most productive when 30-50 years old, and can reach ages of 200 years old. They are planted in lines with a regular spacing that reflects the local rainfall availability: closer in the north of the Sahel and further apart in the south where annual rainfall is lower. The majority of olive trees are found in small holdings with less than 100 trees each, but some large farms have up to 6000 trees.

Traditionally olive cultivation and harvesting is a community activity. Olives are picked by hand in the months of December, January and February. Men stand on A-frame ladders set within the olive trees and claw at the branches with fingers protected by hollow sheep's horns. Below, the women spread sacking, blankets or plastic sheets on the ground and collect the olives that fall from the tree above. They then separate the olives from the leaves and use the blankets to load the olives onto a cart. This picture does traditionally have a strong gender distinction:

men in the trees, women on the ground. This is often now blurred. Women and men mix their traditional roles and more often than not it is the women who stand on the ladders set in the olive trees.

At the start of the French protectorate in 1881 it has been estimated that Tunisia had about 400,000 olive trees. This number increased to over 20 million trees in the 1950s and over 50 million trees in the 1990s. Olive tree cultivation now covers approximately one third of Tunisia's agricultural land and occupies about a quarter of a million growers for at least part of the year.

Olive oil processing plants used to be very common in the Sahel. The traditional method of producing oil from olives is to build circular layers of straw mats and olives approximately one metre in diameter and then squeeze the sandwiched layers from the top to press out the olive oil. The sandwich typically consists of more than 30 alternating layers of mats and olives. The power for this cold pressing technique was traditionally provided by donkeys, but most olive oil plants now run on electricity. During the 1950s there were over 700 olive oil processing plants in the Sahel alone (Despois 1955) and each town and village typically had several olive oil processing plants. Today the number of plants has reduced through consolidation even though the number of trees and the production of olive oil has increased. Southern Sousse has a large olive oil processing plant noticeable by its large chimney and (during December to February) its strong smell.

Olive oil is a very significant part of Tunisia's agricultural production and of the country's agricultural exports. With 160,000 tons of olive oil produced each year Tunisia is the fourth largest olive oil producer in the world. It exports about 70 per cent of its production, mainly to Europe, making it the second largest olive oil exporter in the world. Olive oil accounts for over 40 per cent of Tunisia's agricultural exports and 4 per cent of all Tunisia's exports.

While olive tree cultivation and olive oil production have increased, not least because of the dietary benefits of olive oil, the Tunisian olive oil sector has certain weaknesses.

The equipment used both for harvesting and for olive oil production is often old, simple craft equipment.
Harvesting is carried out predominantly by hand with little mechanisation.
There is little training available to workers to improve olive oil quality.
There are no uniform standards of olive and olive oil quality.
There is no perception in the export market of a unique Tunisian brand of olive oil, as there is, for example, with Italian extra virgin olive oil, and so there is no build up of brand loyalty.

Tunisian government policy is encouraging improvements in olive oil quality and in developing a *Carthage* brand for Tunisian olive oil, a brand name that markets continuity for a product that in some European countries was a luxury until recently.

Change

While olives represent continuity in the Tunisian Sahel, there has been extensive change in the agriculture sector during the last 120 years. During the first ten years of the French protectorate (1881-1891) 93 per cent of Tunisia's agricultural land was in the hands of 16 French land owners. This figure did reduce during the French period, and by independence in 1956 French land owners held 18 per cent of Tunisia's agricultural land, although this land produced 40 per cent of total agricultural production.

In the ten or so years after independence there was left wing pressure on President Habib Bourguiba for land reform. In 1956-1958 agricultural plots above 50 hectares (that is the relatively large plots) were expropriated by the government and redistributed to landless labourers who were required to repay the cost of the land over 25 years. This land reform was concentrated on the lower Medjerda Valley in the north of Tunisia, and the Sahel region was largely neglected.

The year 1969 was a year of agricultural revolution. In January all agricultural land was nationalised in cooperatives as a response to left wing social and economic models practiced elsewhere in the world. However, farmers found the cooperatives unworkable and in September the government agreed that private holdings could be withdrawn from the cooperatives, and in December the cooperatives were largely abandoned with only 10 per cent of land remaining in cooperatives. So, 1969 had swung from complete nationalisation back to a position similar to that at the time of independence in 1956.

Since 1969 the Tunisian government has been active in investing in agriculture to promote improvements through grants and infrastructure investments, but has not nationalised land holdings. The following three examples illustrate government investments in the agriculture sector as part of national development plans.

During the Fifth Economic Development Plan (1977-1981) TD 800 million was invested in agriculture of which TD 250 million was spent on developing agricultural infrastructure.

During the Eighth Economic Development Plan (1991-1996) there was government investment of 2000 million TD in agriculture and policies were developed to encourage private land holdings and private investments.

During the Ninth Economic Development Plan (1997-2001) there was 1770 million TD invested in irrigation, 570 million TD in farm equipment and 230 million TD in water and soil conservation projects.

These investments are part of the process of improving the agriculture sector and meeting the national objective of greater integration in the global economy. Much of the investment combats problems created by the physical environment and particularly the need for careful management of water resources.

Agriculture and water

One significant form of agriculture investment has been in water resources, especially valuable to the Sahel region with only 300 mm rainfall a year. The Nebhana reservoir and dam (see Chapter 2) was one of the first large scale dam projects in Tunisia. The Nebhana reservoir, completed in 1969, holds 85 million m^3 of water and is capable of providing 4000 m^3 of water per second at its maximum rate. The water is then distributed through a large distribution network. Figure 11.1 shows the water distribution network for the coastal part of the Sahel region. Much of the water is destined for areas targeted for agriculture, the irrigated perimeters, where the water is distributed through a network of pipes of reducing sizes.

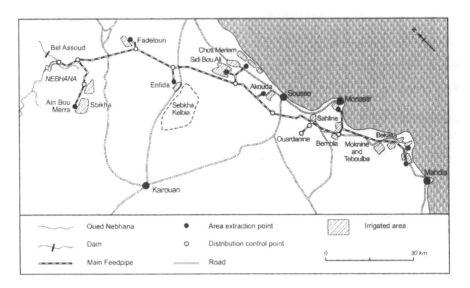

Figure 11.1   The Nebhana water distribution network

Several different irrigation techniques are used in the irrigated perimeters and elsewhere. The following list gives the commonly-used techniques in ascending order of cost.

Earth, clay and plastic open channels
Buried clay pipes
Metal or plastic pipes placed on the surface of the ground
Drip irrigation
Sprinkler irrigation.

The main irrigated perimeters in the Sahel in which these irrigation techniques are used can be found near Teboulba, Bekalta, Sahline, Chott Meriem, Akouda and Sidi bou Ali. The area covered by each irrigated perimeter is approximately 2000–3000 hectares with field sizes of the order of 2-5 hectares. Figure 11.2 is a map of the irrigated perimeter at Sidi bou Ali, north of Sousse. The map shows the areas of irrigated agriculture, those areas not suitable for irrigation and the water distribution network. The area has designated regions with water supplied through major and subsidiary pipes to the fields.

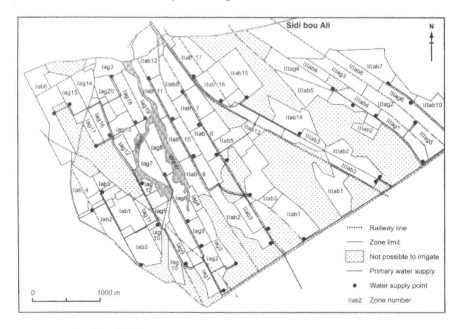

Figure 11.2   The Sidi bou Ali irrigated perimeter

Figure 11.3   A broken water pipe losing water in the Bekalta area

The irrigation techniques used in the Sahel do create significant water losses. Evaporation of open water, leaks from pipes, seepage in sandy soils and poor control of water all contribute to water losses between the reservoir and the farm. Figure 11.3 shows an example of water losses from a metal distribution pipe. In addition, when water prices are low the discipline to control water use carefully can be lost.

Irrigation water is used on high value crops. During the 1970s and 1980s there was much investment in plastic cloches. These cloches are long plastic tunnels to protect the crops and are either about 20cm or about 3m high, supported internally by metal frames. They are used to grow high value crops such as tomatoes, peppers and courgettes in the larger cloches and strawberries in the smaller ones. In some areas they have replaced traditional olive cultivation, and in other areas they have allowed new land to be used for farming. However, the plastic cloches require maintenance of the plastic, which is expensive, and the crops need chemical spray treatment to ward off plant diseases that otherwise flourish in the warm, damp environment of a cloche. On these two counts their sustainability is questionable, and in the Sahel there is much evidence of cloches falling into disrepair.

Box 11.1 Agriculture and irrigation around Sahline and Sidi bou Ali

Around two thirds of the fields we mapped in the Sahline irrigated perimeter in November 2002 were irrigated, and of those that were not around half remained fallow. Potatoes were frequently drip irrigated and tomatoes were often grown in cloches. Sahline also has a large rose nursery which could not exist without irrigation water. Irrigation is important in crop diversification because it allows a larger range of crops to be grown. In both Sahline and Sidi bou Ali we saw irrigated crops such as peppers grown under olive trees. This technique allows a farmer's land to be far more productive than it was previously when it just grew olive trees.

Irrigation systems appear to be vital for crop diversification and in the rural economy of the Tunisian Sahel more generally, but are they sustainable? When we visited Sidi bou Ali we observed that many of the water pipes and pumps of the irrigation system did not appear to be working and the area was littered with broken plastic from the water pipes and the cloches. When fields were irrigated the drip method was the most common type of irrigation used, although there was a large proportion of fallow fields. In Sahline on the other hand the irrigation systems appeared to be both efficient and well maintained. Replacement plastic covers for the cloches are expensive, but in Sahline cloches seemed to be a popular farming method and were in good condition and used effectively.

Adapted from the work of Fiona Howie and Christine Kogler, UCL Geography undergraduates, November 2002

During the 1990s some irrigated areas formerly devoted to agriculture under plastic cloches have been changed to open fields irrigated by drip irrigation from small bore plastic pipes. The crops are of relatively high value such as early season new potatoes. The land itself needs careful attention to take out stones and weeds, but this open approach to agriculture is more sustainable, even though there is still a clear dependence on water. As noted in Chapter 2, this dependence on water in the Tunisian Sahel will increase as rainfall decreases and temperatures increase with predicted climate change.

The picture of change in agriculture in Tunisia also applies to infrastructure investment. Offices to provide advice on crop cultivation, marketing agricultural produce and on water supplies have been established in many large villages such as Bekalta and Teboulba in the Sahel. One such initiative is the offices of the OMIVAN organisation – Office de Mise en Valeur d'Eau Nebhana – that provides information on irrigation techniques, irrigation timing and on the cost of water.

Other examples of infrastructure investment are electricity distribution, access to safe drinking water and transport. In rural areas access to electricity has increased from 76 per cent of the rural population in 1996 to 87 per cent in 2001, and access to safe drinking water for rural peoples has increased from 68 per cent to 87 per cent during the same period. The area from Sousse to Mahdia has seen a new metro rail system that has opened up links to the large villages in the area and resulted in substantial village expansion.

Fishing

Tunisia is blessed with 1300 km of coastline bordering wide continental shelf waters with the warm, shallow waters of the Mediterranean Sea. The periodic scirocco winds that blow from the Sahara desert to the south bring dust and sand deposits (Prospero 1999) that are rich in nitrogenous salts and phosphorous which act as a fertiliser. The conditions are very favourable for fishing and Tunisia has built up a significant fishing sector, at least on the small vessel scale. In 1998 the fishing sector brought in revenues of 126 million TD and provided direct and indirect employment to 100,000 people. The Sahel towns of Sousse, Monastir, Mahdia and Sfax are all prominent fishing ports in Tunisia, with Mahdia a traditionally important location for fishing.

In 1974 national fish production in Tunisia was 43,000 tonnes. By the 1990s the annual catch had increased to around 87,000 tonnes per annum (see Table 11.1), a level it has stopped at despite national plans to increase annual production to 99,000 tonnes per annum during the Ninth Economic Development Plan. Within the total catch there are three main types of fishing: coastal, peche au feu and trawling (Table 11.1).

Coastal fishing is characterised by small boats using nets in shallow coastal waters. It accounts for the majority of Tunisian fishing vessels, and in 1997 over one third of the total catch.

Peche au feu is fishing at night with lights to attract the fish. Four boats cooperate in a small region of the sea with their lights switched on to attract the fish. Three boats switch off their lights to leave the lampero boat with its single light around which the fish congregate. The three remaining boats close in on the lampero with their nets to catch the fish. This type of fishing is common throughout the Mediterranean, mainly in the period from April to October. In 1997 peche au feu accounted for one third of the national Tunisian fish catch. Trawlers operate outside the three mile coastal zone and have large nets for deep sea operations. In 1997 trawling accounted for a quarter of the national catch.

Table 11.1 Tunisian fish catches in tonnes in 1990, 1995 and 1997

Category	1990	1995	1997
Coastal fishing	40,511	27,626	31,497
Peche au feu	26,779	33,816	30,803
Trawler fishing	17,473	17,518	22,092
Other	3,850	4,642	4,634
Total	88,613	83,636	89,026

Source: UTAP (2002)

Of the national catch of around 87,000 tonnes per annum exports account for about 15,000 tonnes a year, although during the 1990s there were considerable fluctuations (see Figure 11.4). Table 11.2 shows the main contributors to the export total for the years 1990, 1995 and 1997. The main countries that receive Tunisia's fish exports are Italy, France, Spain, Germany, Greece, Belgium, the United Kingdom and Japan. Japan takes all of the tuna catch of Tunisia.

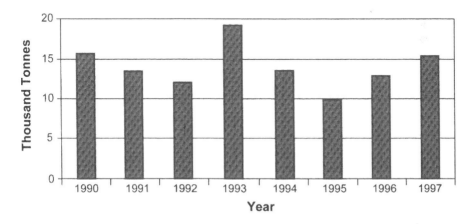

Figures 11.4 Tunisia's national fish exports, 1990–1997

Table 11.2 The main categories of Tunisian fish exports (Data are in tonnes)

Category	1990	1995	1997
Fresh fish, refrigerated or frozen	1,940	1,336	1,506
Canned fish	206	44	274
Prawns or other shellfish	2,732	3,200	4,319
Other sea food	3,003	1,000	181

Source: UTAP (2002)

In the Sahel one of the largest fishing ports is Mahdia. Fishing has been important in Mahdia since the Aghlabids moved their capital there from Kairouan in the 10th century AD. The fishing vessels are mainly the coastal and peche au feu types, and regularly 1400 boats use the Mahdia port. In 1958 the port was reconstructed, including a large covered area to prevent the fish recently caught from drying out too quickly. Since the 1970s the Japanese government has provided aid to support fishing in Tunisia, and in the year 2000 provided 9 million TD for a fisheries training centre in Mahdia. Fish factories have grown up around Mahdia specialising in canning, fish freezing and the production of animal feed. These factories provide over 6,000 jobs.

Because fish resources are under-exploited in Tunisia, particularly in the northern sea regions, government financial support is available to assist the fishing sector. The support takes the form of tax incentives, suspension of value added tax for equipment made in Tunisia, finance for improving fishing vessels, support for fishermen taking educational courses and subsidies on fuel for boats.

Box 11.2 Tunisian government policy on agriculture and fisheries

At the 12th Congress of the Tunisian Agriculture and Fisheries Union, 6-8 June 2000, President ben Ali summarised his government's support to the agriculture sector. The main items are listed below.

Agricultural output had increased by 40 per cent since The Change of 1987.

Water is the core of the sustainable development programme, with 2 billion TD invested in dams and mountain lakes.

Improvements have been made for agricultural workers: increases in the coverage areas of electricity and safe drinking water; improved rural health care and education; better housing; new communication routes.

In times of low rainfall the government has provided support to buy feedstuffs for cattle and has rescheduled debts in the areas badly affected.

Education improvements have included agricultural training, agricultural extension services and targeted scientific research.

Chapter 12

Islam in the Sahel

Introduction

Islam creates perhaps the most obvious visible differences between Tunisia and western Europe. Mosques, minarets, the call to prayer and the dress of women are dramatically different features, particularly in the urban landscape, when compared to the landscapes of Britain, France or Germany. On a world scale, it is arguable that Islam produces greater differences compared to western Europe than any other major culture: Europeans have closer cultural affinities to the United States, Australia, Brazil and Argentina than to Tunisia, Libya and Algeria. And yet, the Islamic countries of North Africa are the closest geographical neighbours to western Europe, with Tunis only two and a half hours flying time from London and two hours from Paris. The attacks in New York and Washington DC on 11 September 2001 have heightened the relative lack of knowledge about Islam in western countries and have promoted to a certain extent greater inquiry about Islam.

This chapter first introduces the religion of Islam, then focuses on the architecture of mosques, and finally looks at Islam in the Sahel cities of Sousse and Kairouan. Islam provides a strong theme of continuity in Tunisia: since the 7th century AD Islam has been the religion of most of the inhabitants of Tunisia and many of the major buildings and customs date from the early years of Islam in North Africa.

Islam and Mohammad

The principal characteristic of Islam is submission to God. The God of Islam is the same God as in Judaism and Christianity in that all three major religions worship a single God. In Arabic the name of God is Allah, which can be dissected as Al Lah and translated as The Being. Submission to God or Allah means accepting the will of Allah, and so a follower of Islam, termed a muslim, is one who submits to the will of Allah.

The founder of Islam was the prophet Mohammad. He was born in about the year 570 AD and raised by his uncle, Abu Talib of the Qureysh tribe, in what is now the western part of Saudi Arabia. At the age of 25 years Mohammad married his wife Khadijah. Mohammad became dissatisfied with the belief in many gods, polytheism, that characterised the tribes of the western Arabian peninsula and he came to believe in a single God, Allah.

At the age of 40 years (in 610 AD) Mohammad received revelations from the angel Gabriel on Mount Hira near Mecca. The revelation or instruction by the angel Gabriel was for Mohammad to 'Read' even though Mohammad was illiterate. Mohammad wrote down the revelations he received from the angel Gabriel in the Koran (also transliterated as Qur'an) which means 'The Reading'. Mohammad continued to have revelations throughout the rest of his life, all of which were written down and make up the 114 chapters or surahs of the Koran.

Mohammad used the revelations for his preaching in the area around Mecca. He converted some of the tribal peoples to the new monotheistic religion he was preaching, including Khadijah his wife, and his followers became known as muslims. However, his preaching did result in persecution and attacks from local tribes, and in 622 AD Mohammad fled from Mecca to Medina (then called Yathrib) some 400 km to the north. This flight, known as the *Hijrah*, is important because it defines the beginning of the Islamic calendar, and explains why in Islamic countries two dates are commonly used: the date used in western countries (Anno Domini – AD) and the date after the Hijrah (AH).

By persuasion and conversion Mohammad became the ruler of Medina. He and his followers fought battles with the tribes of Mecca, including the Qureysh tribe, of whom his uncle Abu Talib was a member, and with tribes from Syria. In 630 AD Mohammad and his supporters marched on and conquered Mecca. By 631 AD Mohammad had become the ruler of Mecca and effectively the ruler of Arabia. He died in 632 AD in Medina.

After his death the revelations that Mohammad had written down in the Koran and subsequently preached were disseminated both widely and rapidly. In a little over 100 years after his death his supporters had spread the teachings of the Koran, that is the religion of Islam, as far as the Atlantic cost of Spain in the west through to the Indus River in Afghanistan in the east. Figure 12.1 shows the spread of Islam after the death of Mohammad, and Table 12.1 gives the corresponding seats of power of the muslim caliphs. The spread of Islam happened at remarkable speed; for example, much of Spain was muslim by 750 AD and stayed muslim until the fall of Granada in 1492.

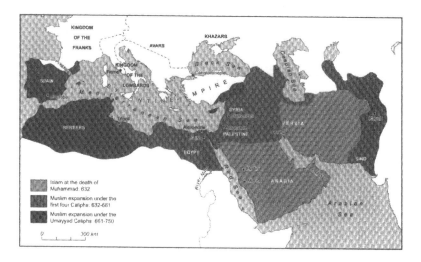

Figure 12.1 The spread of Islam

Table 12.1 The expansion of Islam and the seats of power

Name	Seat of power	Period
Orthodox Caliphate	Mecca and Medina	632 - 661 AD
Ommayad Caliphate	Damascus	661 - 750 AD
Abbasid Caliphate	Baghdad	750 - 1258 AD
Ottoman Empire	Constantinople	1350 - 1918 AD

Five Pillars of Islam

As a set of revelations the Koran is the word of Allah. It is believed by muslims to be infallible and without error, and it has remained unchanged since it was written down by Mohammad during his lifetime. In mosques the decoration on the walls is often a selection of verses from the Koran and so when one looks at these decorations one is looking at the direct word of Allah. The surahs or chapters of the Koran were revealed in Arabic and this language remains the language of the Koran. The verses of the Koran are learned and recited by muslims throughout the world in Arabic irrespective of the local, vernacular language of the believer. So, muslims in Malaysia, Nigeria and Britain know the Koran in Arabic.

Unlike Christianity, Islam has no priesthood and no sacraments. The religious men in the mosques are imams or holy men who act as leaders rather than priests. Muslims follow the Koran as a guide to life: the Tawhid defines what a muslim should believe and the Shari'a defines what a muslim should do. Hence the reference in the west to Shari'a law in some muslim countries. In Islam there are two broad sects:

the Sunnis are the direct followers of Mohammad and make up the majority of muslims, including those in Tunisia;
the Shi'ites follow a line of succession from Mohammad through his son-in-law Ali, and are the majority in Iran.

Islam has five pillars that provide both the fundamentals of the faith and a practical guide to living. The five pillars of Islam are outlined below, taken largely from the web site *Understanding Islam* (Anon 2002).

Shahada

There is no god worthy of worship except Allah, and Mohammad is His messenger. In Arabic the first part is *la ilaha illa 'Llah* – there is no god except God. Ilaha (god) can refer to anything which we may be tempted to put in place of God, for example wealth or power. Then comes *illa 'Llah*: accept God, the source of all Creation. The second part of the Shahada is *Muhammadun rasulu 'Llah*: Muhammad is the messenger of God.

Prayer

One of the icons of Islam commonly used in the west is the call to prayer sung by a muezzin from a minaret. When radio or television programme producers want to give an immediate impression of Islam they typically use the call to prayer, or at least the first part – Allahu akbar. The beginning of the call to prayer is (in English):

> God is most great. God is most great.
> God is most great. God is most great.
> I testify that there is no god except God.
> I testify that there is no god except God.
> I testify that Muhammad is the messenger of God.
> I testify that Muhammad is the messenger of God.
> Come to prayer! Come to prayer!
> Come to success (in this life and the Hereafter)! Come to success!
> God is most great. God is most great.
> There is no god except God.

Muslims pray five times a day: at dawn, shortly after noon, mid-afternoon, sunset and early night. Because these broad times vary during the year with the changing rising and setting of the sun the actual prayer times vary on a day-to-day basis. Table 12.2 lists the actual prayer times for Sousse for 1 August 2002. The prayers are verses from the Koran, and so are the direct word of Allah. It is not compulsory to attend a mosque in order to participate in the prayers, although it is normal to face in the direction of the Ka'bah or black stone which is located in the main mosque in Mecca and which Allah commanded Abraham and Ishmael to build 4000 years ago.

Table 12.2 Prayer times for Sousse, 1 August 2002

Dawn	Shortly after noon	Mid afternoon	Sunset	Early night
Fajr (start-end)	Thuhr	Asr	Maghrib	Isha
03.30 - 05.19	12.32	16.19	19.28	21.03

Source: http://www.namazvakti.com/

Self-purification

One of the most important principles of Islam is that all things belong to God, and wealth is therefore held by human beings in trust. In a geographical context the environment belongs to God and is held in trust by humans (Izzi Dien 2000). This pillar is known as *zakat*, a word that means both 'purification' and 'growth'.

Box 12.1 Christianity and the Koran

Many of the themes of Christianity are held in common with Islam. There is a
striking similarity between the first surah of the Koran and the Lord's prayer in
Christianity. Both are reproduced here for comparison.

The Lord's prayer   The Koran Surah I The Opening

Our Father who art in Heaven   Praise be to Allah, Lord of the Worlds,
Hallowed be thy name   The Benificent, the Merciful
Thy kingdom come   Owner of the Day of Judgement,
Thy will be done   Thee (alone) we worship; Thee (alone)
On Earth as it is in heaven   we ask for help
Give us this day our daily bread   Show us the straight path
And forgive us our trespasses   The path of those whom thou hast
As we forgive those who trespass   favoured;
  against us   Not (the path) of those who earn thine
And lead us not into temptation   anger
But deliver us from evil   nor of those who go astray

Fasting

Every year in the month of Ramadan, the ninth month of the lunar calendar,
muslims fast from dawn until sunset, abstaining from food, drink and sexual
relations. Those who are sick, elderly or on a journey, and women who are
pregnant or nursing are permitted to break the fast and make up an equal number of
days of fasting later in the year. If they are physically unable to do this, they must
feed a needy person for every day missed. Although the fast is beneficial to health,
it is regarded principally as a method of self-purification. The end of the month of
Ramadan is the celebration of Id al Fitr.

Pilgrimage

Muslims are expected to undertake the pilgrimage to Mecca – the Hajj – once in
their lifetime if they are physically and financially able to perform it. About three
million people go to Mecca in Saudi Arabia each year from every part of the world.
The rites of the Hajj, which are of Abrahamic origin, include circling the Ka'bah
seven times, and going seven times between the mountains of Safa and Marwa
thereby reflecting the travels of Hagar during her search for water. The close of the
Hajj is marked by a festival, the Id al Adha. The Id al Adha and the Id al Fitr are
the main festivals of the muslim calendar.

Islam and Learning

Since its early years Islam has been associated with knowledge. The university of Al Azhar in Cairo and the Al Zaytunah university in Tunis are the oldest existing universities in the world, founded some 1000 years ago. Islamic science was in its prime during the period from the ninth to the thirteenth century AD, that is the time of the European Middle Ages. Many scientific terms and names have Arabic origins, such as alcohol, carat, chemistry, magazine, sugar and zenith (Hattstein and Delius 2000), as well as place names such as Gibraltar, Al Hambra and Almeria.

In astronomy many star names have Arabic origins such as Algol, Deneb and Betelgeuse, while European scientists including Copernicus, Tycho Brahe and Kepler all used muslim astronomical tables in their work. Many of the astronomical tables were calculated by the ruler Ulugh Beg of Samarkand (1394-1444) who drew up the most precise astronomical charts of the Middle Ages. Islamic astronomers calculated the courses of the stars, the dimensions of the Earth and predicted the weather and the state of water supplies. Al Hazen (965 – died after 1040), for example, calculated the amount of water in the River Nile for agricultural purposes.

In mathematics muslim and Arabic scholars gave us the base 10 decimal system and Arabic numerals, and made substantial contributions to algebra, the development of algorithms and trigonometry. The numbers used in the west and those used in most Arab countries (see Table 12.3) appear to be distinctly different. Curiously, the west uses Arabic numbers and the Arab world uses numbers that have an origin further east. This simple picture may conceal more overlap than is at first apparent from Table 12.3. The Arabic numbers for 1, 5, 6 and 9 are used in a similar form in the west, although not necessarily directly for the same numbers. The Arabic number for 4 can be reversed to make a western 3, while the Arabic numbers for 2 and 3 can be turned on their side to make the same numbers used in the west, albeit with long tails in their Arabic versions.

Table 12.3 Numbers 1–10 used in the west and in the Arab world

1	2	3	4	5	6	7	8	9	10
١	٢	٣	٤	٥	٦	٧	٨	٩	١٠

In medicine muslim scholars built on the work of the ancient Greeks and there was a substantial improvement in the understanding of pharmacology, infectious diseases and therapeutics, and above all the treatment of eye disorders. Perhaps the most famous physician of this period was Ibn Sina of Bukhara (980-1037, see Figure 12.2), known in the west as Avicenna. He was a polymath who made contributions to philosophy, astronomy, grammar, poetry and medicine. His compendium of medicine was regarded as a standard work in Islamic and European countries for several centuries after his death.

| Al Idrisi | Ibn Khaldun |
| Ibn Battuta | Ibn Sina |

Figure 12.2 Four of the major Islamic scientists of the Middle Ages

Some of the earliest explorers and map makers were muslims; it is noteworthy that the Koran encourages exploration and concern for the environment (Izzi Dien 2000). Three of the most notable muslim explorers are described below. Sketches of what they are reported to have looked like are shown in Figure 12.2.

Ibn Khaldun (1332-1395 AD) was born in Tunis of parents who came from Yemen via Spain. He moved to Morocco and then to Algeria before spending most of his life in Egypt where he was the Chief Malakite judge and also lectured at the Al Azhar University. Ibn Khaldun's first main contribution to science was a major work on the philosophy of history. He identified the psychological, economic, environmental and social factors that contribute to the development of human civilisation. He analysed the dynamics of group relationships and showed that new civilisations and systems arise from tensions within societies, an early form of the concept of a paradigm shift (Kuhn 1996). Ibn Khaldun's writings on world history deal with the Arabs, European rulers, Jews, Greeks, Romans, Persians, ancient Egypt and North Africa. His volume on North Africa is largely autobiographical and began a tradition of analytical approaches to biography.

Ibn Battuta (1304-1369 AD) was born in Tangier and began his extensive travels at the age of 21 years. He is the only muslim traveller of his period known to have visited all the muslim countries of the time; his journeys covered a total of over 75,000 miles. He travelled extensively in the Middle East, including several visits to Mecca for the Hajj, India, China, the Far East, Africa, Spain and Russia. He visited China only 60 years after Marco Polo. His writings are compilations of his travel experiences and include descriptions of physical geography, economic geography, agriculture, food, religious customs and social life. He shows that muslim sailors dominated maritime activity in the Red Sea, the Arabian Sea, the Arabian Gulf, the Indian Ocean and in Chinese waters during the 14th century AD. Ibn Battuta now has a crater on the Moon named after him.

Box 12.2 Islam, urbanism and architecture

The Arab world has a history of urbanism, and Islam has played an important role in determining urban form and function. But how is Islam expressed through modern buildings? What influence does Islam have on modern urban architecture in comparison with traditional Islamic architecture found within the medina walls?

Hillenbrand (1983) points out three basic categories that distinguish Islamic architecture. First, separate buildings are juxtaposed within an overall entity. This illustrates the Islamic concept of the urban area resembling an organism, and does not prioritise areas or functions. This concept is manifested through the rows of plain, white walls and narrow, convoluted streets of the medinas. Islam discourages excessive expenditure on buildings. The exterior should give no indication of social status. Consequently, urban districts and buildings are inward looking, and the inner space is sanctified by shuttered windows so that owners can see out, yet the public cannot see in (de Montequin 1983).

Second, a few standard architectural elements are continually re-used and regrouped within single buildings as domes and arcades. Al Bayati (1984) claims that Islamic architects prefer to refine existing designs rather than experiment with new forms, with interchangeability being a key aspect of Islamic architecture. Hillenbrand (1983) notes that Islamic architects use features such as symmetry, receding vistas and axiality.

Third, Islam declares that man should not attempt pictorial representations of reality because Allah is the sole creator of life (Bianca 2000). Symbolic representation is an appropriate way of dealing with this principle, the main tools being calligraphy, geometric patterns and arabesque art.

Colours play a significant role in Islamic architecture. White, blue and green are used often and have symbolic attributes. White represents purity and is used on external walls. Blue symbolises good luck and is associated with heaven and with Allah. Green symbolises human and environmental fertility.

Adapted from the work of Camilla Flatt, UCL Geography undergraduate, November 2002.

Al Idrisi (1099-1166 AD) was born in Ceuta and educated in Cordoba. His major contribution lies in the cataloguing of medicinal plants and the development of new drugs. He compiled a geographical encyclopaedia containing material on the physical geography, economic geography and cultural geography of Asia, Africa and Europe. This encyclopaedia was delivered to the Norman King Roger II at his court in Palermo, Sicily. The book had the title *Al Kitab Al Rujari* (Roger's Book) plus the sub-title *The delight of him who desires to journey through the climates.* Later on Al Idrisi compiled another encyclopaedia, larger than Al Kitab Al Rujari, which included material on geography, botany, fauna and zoology. His work was translated into Latin and was popular in Islamic and western countries for several centuries after his death.

Mosque Architecture

This section summarises some of the main characteristics of mosque architecture, although as we shall see in the following sections, details vary from mosque to mosque, even within a single country.

A useful way to understand the architecture of most mosques is to start from the inside and work outwards. The inner sanctum of all mosques is the prayer hall – where non-Muslims are usually not allowed to enter at all and to enter which Muslims must have ritually washed themselves (and women must cover their heads). Worshippers face the direction of Mecca when they pray, and the *mihrab* (or alcove) indicates the direction of prayer in the prayer hall. The *mihrab* in turn is located in the *qibla* or facing wall. It is important, however, to emphasise that it is not the *mihrab* or *qibla*, but the direction they indicate, that is sacred.

Prayer halls are normally broad rather than long (unlike most churches). Their decoration varies from the very plain to the fabulously ornate (often with superb chandeliers), but none contain any representations of living beings. There is very little religious furniture. In most prayer halls there is a *minbar*, which is the pulpit for the *imam* or preacher, a *dikka* (small platform) from which a *muezzin* would traditionally have amplified the imam's voice, and a kursi or lectern. There are no pews in prayer halls; worshippers sit on the floor which is usually covered with rugs or rush matting.

Mosques are normally constructed around the axis passing through the *mihrab* and at a right angle to the *qibla*. The prayer hall, as it is ritually pure, is separated from the rest of the mosque by a step or balustrade. Outside, the principal other part of the mosque is the courtyard. This has multiple functions – it can be used as a spillover for worshippers when the prayer hall is full, but it is also used for communal activities. In some mosques the courtyard is surrounded by a colonnaded area, which is often used for shelter.

The mosque as a whole is in turn separated from the outside world, usually with thick, windowless walls and impressive gates. It may be physically separated from the rest of the medina, but the influence of the mosque nevertheless extends beyond its walls. It does so in a literal sense – the call for prayer can be heard throughout the medina. But also in a morphological sense – as we shall see in the next chapter mosques traditionally formed the centre of medinas, and a symbolic sense too – the mosque is the centre of Islam in the medina.

Islam in Tunisia

Sousse

Islam is the official religion of Tunisia and 98 per cent of the population are muslim. Tunisia has some of the oldest Islamic buildings in the world and this section will concentrate on the Islamic buildings in Sousse and Kairouan.

In Sousse there are two notable 9th century Islamic buildings, the Grand Mosque (Figure 12.3) and the ribat (Figure 12.4). The Grand Mosque was built in 851 AD by Emir Muhammad I and is in the traditional style of mosques found in the Arabian heartland, North Africa and Spain (Frishman and Khan 1994). It has a main courtyard flowing out from the prayer hall and defensive walls on which verses from the Koran are carved in *bas relief.* The mosque has no minaret in the strict sense, but rather an 11th century kiosk in the north west corner. The architecture of the mosque is exceptionally simple and the building has no decoration other than Koranic verses. The prayer hall has 13 aisles and is aligned so that muslims can pray facing the direction of the Ka'bah in Mecca, also termed the Qiblah direction or the direction of the end wall of the mosque. There are many stone pillars in the three bays of the prayer hall, in part a reminder of the trees in the natural world that are owned by Allah and held in trust by human beings. The Grand Mosque in Sousse is today used by about 300 worshippers for the main Friday prayers shortly after noon (the Asr prayers), and 60-80 worshippers at other prayer times during the day.

Figure 12.3   The Grand Mosque in Sousse

The ribat in Sousse (see Figure 12.4) was constructed in 821 AD and is a combination of a fortress and a religious building. It was originally built as a lookout tower, one of a string of North African towers situated on the coast that could send a message by light and fires from Alexandria in Egypt to Ceuta in Morocco in one night. While the original purpose was a defensive one, the building was also used for study and prayer. The ground and first floors of the ribat have cells that were used for prayer and for teaching the Koran to small groups clustered round an imam. The ribat brings together the concepts of meditation and holy war or *jihad*: peace and seclusion for followers of Allah plus a meeting place for the warriors of the holy war. *Jihad* however is wide ranging in its meaning, and may apply to war against poverty and disease or an intellectual or spiritual struggle as much as it applies to battles between armies or individuals. The ribat of Sousse has many architectural similarities with the caravanserais of central Asia and Anatolia: a square ground plan, fortified walls with a round tower at each corner, semi-circular towers in the centre of the curtain walls, rooms grouped around a central courtyard, reservoirs of water and storerooms for food. The ribat in Sousse also has a mosque within its south side.

Figure 12.4   The ribat in Sousse

The Sousse medina has many mosques, the oldest of which is the Bu Fatata mosque built in 840 AD during the reign of Emir Abu Iqbal (Hattstein and Delius 2000). The accompanying minarets reveal themselves down the side streets and alleys of the medina. The minarets are typically either square or octagonal in cross section: square minarets show the influence of the Malekite school of Islam, while the octagonal minarets originate in the Turkish Hanefite school. Throughout many of the mosques of Sousse can be found tile work in blues and yellows that show the links to Andalucia and the Islamic rule of Spain.

Kairouan

Kairouan is said to be the fourth holiest city in the Islamic world after Mecca, Medina and Jerusalem. The Grand Mosque in Kairouan (see Figure 12.5) is one of the oldest in the world and the base of its minaret, dating from 730 AD, makes it the oldest minaret in the world.

Figure 12.5   The grand mosque in Kairouan

Kairouan and its Grand Mosque was founded in 670 AD by Sidi Oqba ibn Nafi, a general of the Omayyad caliph of Baghdad. In its early years the Grand Mosque underwent several reconstructions. It was renovated in 703 AD and then enlarged in 724 AD by the order of the Omayyad caliph. It was renovated once more in 836 AD under Ziyadat Allah (Hattstein and Delius 2000) and the whole mosque we see today (Figure 12.6) is essentially a ninth century construction, except for the base of the minaret.

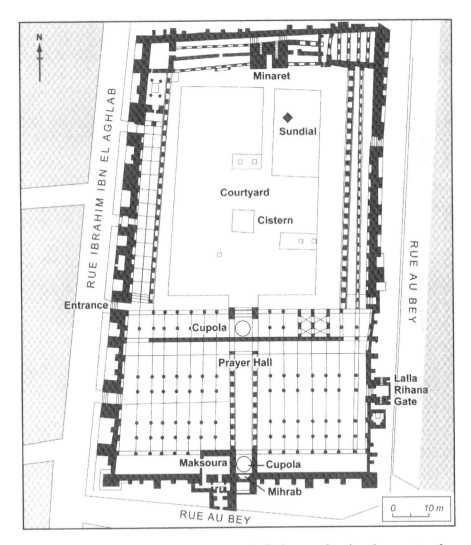

Figure 12.6 **A plan of the grand mosque in Kairouan showing the courtyard, the prayer hall and the minaret**

The prayer hall of the Kairouan Grand Mosque is particularly impressive. It has 414 columns that mainly come from ancient buildings, including the Roman amphitheatre at El Djem (see Chapter 10 of this book). Most of the columns are capped by blocks of olive wood to absorb shocks from the earthquakes that occasionally affect Tunisia. The mosque is simple and elegant. It is a synthesis of the typical architectural forms of religious buildings of the area (see Figure 12.6). The design of the minaret reflects the watch towers of the time of its construction, including the lighthouse tower (the Tour Khalef) of Sousse.

Although today it is only Kairouan that remains as an active town, there were other settlements built nearby during the early centuries of Islamic rule. Al Abbasiya, al Mansuriya and Raqqada were substantial towns built near Kairouan but today are only archaeological sites. Al Mansuriya was built by the ruler al Mansur who came to power in 946 AD. According to al Idrisi the town of al Mansuriya had a circular plan, defensive walls four metres thick, semi-circular and rectangular towers, a large mosque, over 300 baths and the main markets of the region. Al Mansuriya was a high prestige city which may have influenced the development of Cairo, particularly the mosque which has strong similarities with the al Azhar mosque (972 AD) and the al Hakim mosque (990-1013 AD) in Cairo.

Chapter 13

The Challenges of the Sahelian Medinas

Introduction

As we saw in Chapter 3, one of the principal physical manifestations of Islam in the Tunisian Sahel is the medina, or traditional walled urban area. The last chapter, meanwhile, emphasised the speed with which Islam spread, and this has two implications for our understanding of medinas. First, their form and function has been, at least until only very recently, strikingly similar across the Arab World. These are urban areas that evolved almost simultaneously across a wide number of countries dominated in a short time span by Islam. A second implication of the inextricable link between Islam and medinas is that they have a long history – work began on Kairouan's medina as early as 670 AD.

What is very striking is that for over a thousand years the form and function of medinas underwent very little change. The typical medina that is described in the first part of this chapter could more or less have been found throughout the Arab World at any time between the 7th and 19th Centuries. In Tunisia, colonialism by the French heralded the first significant changes in medinas, and change has characterised Tunisian medinas ever since. The second part of the chapter therefore focuses on the evolution of medinas.

Today medinas present a significant challenge to the Tunisian authorities. It is hardly surprising that cities that emerged in the 7th century are not well suited to the demands of the contemporary world. But medinas represent more simply than a planning problem. They are also symbolic of a number of tensions within Tunisian society. They perfectly express the tension between continuity and change. Should 1100 years of history be eradicated to facilitate Tunisia's fuller incorporation in the global economy? Conversely, should medinas be maintained to attract tourists from the industrialised world, even at the expense of Tunisians living within them? The answer to these questions is particularly difficult given the religious connotations of medinas. The final part of this chapter assesses the various responses that have been developed in response to these sorts of challenges.

The Form and Function of Medinas

Most first-time visitors to medinas would characterise them as utterly chaotic, and would probably be surprised to find that traditionally medinas had a very clear structure, much of which is still apparent today, albeit after some orientation.

The morphology of medinas is most straightforwardly depicted as a series of concentric circles (Figure 13.1) (Findlay, 1994). The centre of the medina is the main mosque – usually called the Grand Mosque. The area immediately surrounding the mosque is dominated by broadly religious activities – these would include religious schools, printing, publishing and bookshops. Further concentric circles radiate outwards and contain accommodation and a range of economic activities. The entire medina finally is enclosed by the medina wall.

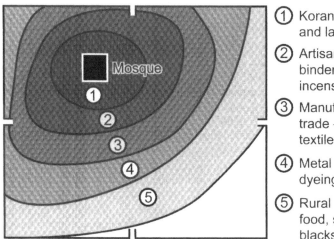

1. Koranic schools and law courts
2. Artisans - book binders and incense sellers
3. Manufacturing and trade - carpets, textiles, carpentry
4. Metal working and dyeing
5. Rural connections - food, saddlers and blacksmiths

Figure 13.1   Concentric circle model of a medina

In many ways this simple model is very useful. First, it emphasises the inextricable link between medinas and Islam – the Grand Mosque is the centre of the medina, and is protected both physically and symbolically by a buffer zone that reinforces the ritual cleanliness and religious purpose of the mosque. The model secondly demonstrates the wide variety of functions that medinas traditionally served. It is important to understand that until the end of the 19th century medinas represented the full extent of urbanisation in Tunisia and across the Arab World. They served all urban functions, including housing, employment, schools, doctors, markets and, of course, religion. At the same time, the model shows the zonal nature of medinas, whereby accommodation and economic activities were largely separate. Third, the model highlights the significance of the medina walls. Gates in these walls would have been locked every night, completely enclosing the medina. A further function of the medina, it follows, was defence.

Appealing though its simplicity is, the model can also distract attention from a series of other important features of the medina. Most prosaically, the Grand Mosque is rarely at the physical centre of the medina, its location often being determined by local physical geography, as we saw of Kairouan's Grand Mosque in Chapter 12. More interestingly, the model does not account for the large number of smaller, neighbourhood mosques that are to be found throughout most medinas. An associated point is that the model does not allow room for analysis within the various concentric circles it depicts. Closer attention, for example, shows that accommodation within medinas is often itself cellular, with neighbourhoods based around local mosques and bathhouses (Toyn and Brierley, 1981). Similarly, economic activities were at least traditionally highly zonal, with a clear physical separation between different economic activities. Finally, although the model does emphasise Islam, it says not much more about society. In fact the structure of medinas often closely reflects society.

In order to try to tease out some of these more subtle issues, it is useful to view the medina at a variety of scales, starting at the micro-level of the individual mosque and house, moving though the scale of neighbourhoods, and finishing at the macro-scale of the overall morphology of the medina. The specifics of the remainder of this section are taken mainly from the medina of Sousse, but apply in almost equal measure to the other medinas both of the Tunisian Sahel and the rest of the Arab World.

Domestic Architecture

The description in the last chapter made clear that mosques have a distinctive structure. Common features, for example, include an outer wall that separates the mosque from the outside world, and a further separation within the mosque between the 'public' courtyard and the 'private' prayer hall. This layered structure is clearly reflected in domestic architecture within medinas.

Like mosques, houses within medinas are overwhelmingly inward looking. Indeed, were it not for one or two museums (for example the Dar Essid in Sousse), it would be virtually impossible to have any idea of what the inside of a house looks like and what goes on there. Characteristics of the exterior of houses are thick, plain and often white washed walls, a striking lack of windows – and where they do exist they are usually grilled or shuttered, and thick wooden doors, often decorated with studs (Figure 13.2). Physical separation between the inside and outside is often further reinforced by a small vestibule immediately inside the front door, where visitors were traditionally asked to wait before finally being admitted across the threshold into the house.

The unwelcoming exterior of these houses, however, contrasts sharply with the inside. Houses were traditionally built around a central, open courtyard, which provide the main communal area within the house. From this courtyard radiate the other rooms. Again in contrast with the exterior, the interior of houses is often very decorative, with colourful and intricate tile work and flowers and plants in the courtyard.

Domestic architecture tells us much about Tunisian society. First, the fact that the morphology of houses mirrors so closely that of mosques is a reminder of just how pervasive Islam is in everyday life. Second, the contrast between the inside and outside speaks volumes about the tensions between the public and private in Tunisia. What goes on inside the house is very private – this was one reason suggested in Chapter 7 why it is so hard to assess the economic activities of women in Tunisia. Outsiders cannot look in, and may not be received in the courtyard at all, instead transacting their business in the entrance hall. Within the house, however, life is very communal and public. Most time is spent in the courtyard, and the main rooms open up onto it. A third aspect of society that can be inferred from the layout of domestic architecture is the tendency, at least until recently, for Tunisians to live within the context of the extended family. Often bedrooms are provided for adults who are not members of the nuclear family.

Figure 13.2   Exterior of a 'typical' medina house

Neighbourhoods, Quarters and Souks

Although they are hard to discern today, traditionally there were neighbourhoods in medinas, with houses typically based around a small family mosque and fountains in open squares for water supply (Toyn and Brierley, 1981). Neighbourhoods usually combined 'similar' people: that is people of a similar 'class' or status, and of a similar ethnicity. In this way a number of neighbourhoods often comprised distinct ethnic quarters within traditional medinas (Abu-Lughod, 1984).

A similar structure was discernible in those parts of the medina devoted to economic activities. Not only were these largely separate from neighbourhoods, in addition within them different types of activity were spatially concentrated. As we shall see in the next section, this pattern has largely disintegrated today, however it is easily retraced through analysis of street names in many medinas. Whole streets were devoted to single industries – ironmongers, leather tanneries, carpet weaving and olivewood carving, for example – taking advantage of economies of scale. These primary industries were in turn separated from more commercial activities, such as the selling of their produce. Even within more commercial areas a distinctive structure emerged. There were different markets for meat and vegetables and fish, and different zones within a more general market selling everything from spices through leather goods to tea and coffee.

Overall Morphology

In reality, then, and even before the changes that are described in the following section, the form and function of most medinas defied the simplicity of the concentric circle model shown in Figure 13.1. That is not to say that there was not a discernible structure within them, but rather that it was more complex, and had greater social implications, than is often depicted.

A more sophisticated model might focus on the 'organic' nature of medinas. They contained distinct areas – for religion, accommodation and employment – each with their own internal, cellular structure. At the same time, these different areas depended upon one another in a relationship that might be characterised as symbiotic. And of course a real strength of the organic analogy is that is allows for change – medinas have evolved over time.

The Evolution of Medinas

Part of the reason it is hard to discern any order in medinas today is simply because orientation is difficult inside them. Streets are narrow and often dark and crowded, for example, and it is often hard to find a viewpoint from which to look down upon the medina and thus form an overall view of its layout. In addition, however, the structures described in the previous chapter have disintegrated in the last hundred years or so, albeit at different rates and to different extents in different medinas.

In the Tunisian context, five key reasons for the changing form and function of medinas can be isolated. The first was large-scale rural-urban migration, which occurred as a consequence of French colonisation after 1881. As described in Chapter 3, one of the impacts of French colonisation was the dispossession of hundreds of thousands of rural landholders, many of whom headed for the city in search of work. The subsequent rise in population density within medinas precipitated a second key moment in the evolution of medinas, which was the relocation outside the walls of French settlers and wealthier Tunisians. A third impetus for change was World War Two, and in particular the bombing of certain Tunisian cities, which in some case destroyed parts of the medina and its walls.

A fourth factor can for the sake of simplicity be termed 'modernisation', particularly of production. New techniques, located outside the medinas and thus not hampered by restrictions upon space, competed with the traditional economic activities within the medina and have largely overwhelmed them. Today the term globalisation might be used instead of modernisation, and there is no doubt that the Tunisian economy as a whole also faces increased competition from overseas. As a related point, the final and most recent dynamic for change in medinas has arguably been the massive take-off of tourism in Tunisia in recent years.

The remainder of this section considers the combined effects of these five factors on the medina. Following the structure of the previous section, it begins at the micro-scale of domestic architecture then works upwards through the scales, concluding with some overall comments.

By the time the rural migrants arrived in the medinas towards the end of the 19th century, there was not much room for expansion there. Their walls fixed the geographical limit of the medinas, and the area within the walls was already largely built-upon. The only possibility to provide accommodation for the new arrivals was therefore to make room available in existing houses. Initially letting out rooms provided a source of income for those who owned houses within the medina. Over time, this pattern cemented itself, and many of the traditional houses were divided into one-room apartments, while the wealthier Tunisians followed the French out of the medina and into the new towns emerging outside the medina walls. Today there are relatively few houses in Sousse medina, for example, that house extended families as they once did. As emphasised in the preceding section, domestic architecture also reinforced Tunisian society, and arguably the decline of traditional houses also indexed a societal change, away from cohabitation with extended families towards more nuclear living.

The overall pauperisation of the medinas also undermined the distinctive nature of neighbourhoods. Whereas once there were distinct wealthy and poor neighbourhoods, now they were all poor. Similarly the boundaries between once-distinct ethnic quarters also became blurred, as the newly arrived migrants took accommodation wherever they could find it, irrespective of ethnicity.

A similar blurring of boundaries has taken place between residential areas and those devoted to economic activity. One way that residential areas have encroached upon areas of economic activity has been through the construction of second and occasionally third storeys above workshops and shops. Mapping the activities at ground floor level and first floor level in the medina present quite different results, with increasing proportions of people apparently living in the same premises as their work.

At the same time, the distinctive concentration of different economic activities is no longer obvious. One reason is that many activities, particularly in production, have now closed down. There are, for example, no jewellers left in Sousse's *Rue des Bijoutiers*. These economic sectors have faced competition initially from within Tunisia but outside the medinas, where modern factories using up-to-date techniques have established significant economies of scale, and more recently from outside Tunisia as its markets have opened up to cheap imports. The most obvious concentration of activities to be found in most Tunisian medinas today is focused on the tourist industry, and the sale of cheap souvenirs, T-shirts and cassettes. Further implications of tourism for medinas are discussed below.

Continuing up the scale, from individual houses through neighbourhoods, quarters and souks, changes are also evident at the macro-scale of the overall morphology of the medina. Most obviously, in many medinas the walls are no longer complete (Sousse is an exception). In Kairouan part of the wall was destroyed during bombing in World War Two. Elsewhere, parts of the wall have been destroyed in order to allow for increased access into, and expansion out of the medina. The next section illustrates this with a case study of Monastir.

An interesting question to ask is what purpose the medina wall now serves. Clearly it is not for defence against attack. Arguably, however, it does still separate the traditional medina from the modern town outside. Changing ideas of security and defence are discussed in greater detail in Chapter 14. Certainly at the macro-scale, one of the most startling changes that has affected medinas is that they now co-exist side by side with modern, largely French towns. The differences between the two areas are often striking. Narrow streets in the medina contrast with wide boulevards outside, and similarly cramped quarters with spacious villas. In general medinas house relatively poor people, and the new towns and suburbs a wider range including the more wealthy. And even life seems more traditional inside the medina. Fewer women are visible than outside, and those who are often are dressed more traditionally.

Which discussion raises a final question – to what extent has the mosque survived changes in the medinas? At one level mosques, and especially the Grand Mosques, represent islands of continuity in a sea of change. None of the major mosques in the Tunisian Sahel were damaged during the war, and all retain their original structure, albeit it with some cosmetic improvements. It is more interesting to ponder the extent to which mosques have maintained their symbolic significance within the medina. If we were to draw a concentric ring model of the contemporary medina, arguably the Grand Mosque would no longer appear at its centre. It would probably be replaced by the main tourist souk, which is where most money is exchanged and where many people – both visitors and locals – spend most of their time (Kairouan would be an exception, where the Grand Mosque is the main reason for tourists to visit). Certainly medinas have a far more secular feel than one imagines they might have had even fifty years ago. Clearly medinas still provide accommodation for some, and employment for dwindling numbers. And the Grand Mosque certainly still has a function during religious holidays. The question is: to what extent has tourism become the main function of medinas today?

Contemporary Responses to the Challenges of the Sahelian Medinas

The preceding section alluded to some of the problems facing contemporary medinas: they have become pauperised, traditional sources of employment have dried up, they are overcrowded, there is little room for expansion, access is extremely restricted and so on.

A range of responses has evolved over time, which might be thought of as existing along a continuum of change. At one end of the continuum is minimal change. This is probably best represented in Tunisia by the medinas of Tunis, Kairouan and Sousse, both of which have been preserved as UNESCO World Heritage sites, one of the stipulations of which status is tight planning restrictions on anything other than cosmetic renewal. In effect these medinas have become historic artefacts. Moving along the continuum, many medinas – examples in Tunisia would be Kairouan and Sfax – show evidence of ad hoc responses to particular problems. Parts of the medina wall have been demolished to increase access, for example, roads have been widened, modern storeys have been added to traditional buildings and so on. Compared with the medinas in Tunis and Sousse those in Kairouan and Sfax feel more vibrant – somehow more alive. They are functioning town centres rather than museum pieces. At the other end of the continuum is large-scale change. This is best represented in Tunisia by Monastir, and the rest of this section focuses on the radical responses adopted there, and on their causes and consequences.

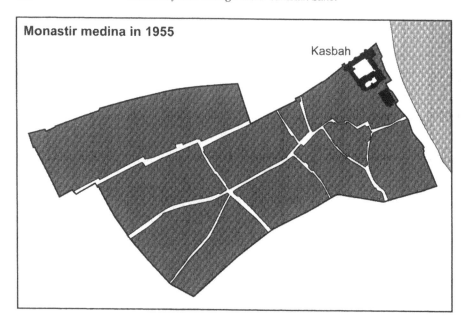

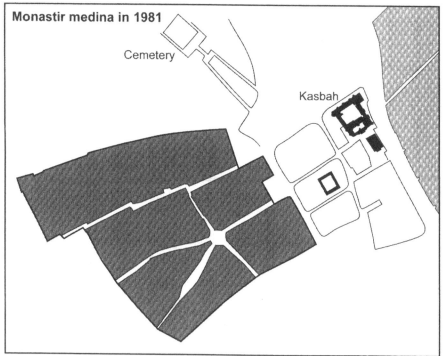

Figure 13.3   The Monastir medina before and after redevelopment

Monastir originally had two separate walled structures – a medina centred on the ribat, and a walled suburb to the west. By the turn of the 20th century these two areas had merged into a single walled medina. In the late 1960s that part of this larger medina that coincided with the original medina around the ribat was demolished (Figure 13.3) – although the ribat itself still stands. What remains of the medina in Monastir today therefore coincides with the original western walled suburb. It is surrounded by wall on three sides – the north, south and west, but not to the east. For visitors this has a strange impact. Walking from the station one enters the medina through a gate in the western wall, and the medina feels much like any other medina in Tunisia. After about 200 metres, however, the medina simply melts away, as narrow roads open into wide boulevards and traditional houses and workshops are replaced by modern monuments.

The area that has been redeveloped focuses on the Bourguiba mausoleum (see Figure 4.1 in Chapter 4). A wide paved pathway stretches about two hundred metres to the palatial style mausoleum where Bourguiba's body now lies. In this redeveloped area also stands the Bourguiba mosque. Otherwise, the area is a rather soulless zone of wide streets and parks that need to be crossed in order to reach the ribat.

Monastir's medina shared many of the problems found in those elsewhere in the Sahel. This was the main reason why the redevelopment took place. It is hard not to conclude, however, that added impetus to the redevelopment came from the fact that Monastir is so inextricably linked with Habib Bourguiba (Chapter 3). What better way to present Tunisia as a modern nation to the outside world than to excise the cramped traditional medina in the President's own hometown and replace it wide boulevards and modern monuments? Perhaps the most symbolic example of this idea in Monastir is the golden statue of Bourguiba as a child, carrying his books to school. The irony is that the statue is built in the middle of a roundabout constructed on the original site of his school.

What have been the consequences of Monastir's redevelopment? Certainly from an aesthetic perspective the result is not pleasing. Monastir's medina is reported to have been one of the jewels of the Sahel: one could stand on the ramparts of the ribat and watch the sea lap at the medina walls below. Now the ribat is isolated from the rest of the old city, and on view from its ramparts to the east is a rather non-descript stretch of road alongside a dull beach, and to the west deserted boulevards and parks. One guidebook goes so far as to suggest that: 'All town planners should be made to visit Monastir, where they would publicly repent the evils of their profession before being put to work rebuilding the old town'.

But more importantly, the redevelopment has inevitably affected the Tunisians who used to live in the medina. In the redeveloped area there is no accommodation and very little economic activity either. Everyone has been displaced outwards into the suburbs of Monastir. And the destruction of the physical infrastructure arguably has gone hand in hand with the destruction of a social infrastructure, as whole neighbourhoods have been dispersed.

On the other side of the balance sheet, there are clearly winners as well as losers. For the few economic activities that are to be found in the redeveloped zone, clearly access has improved dramatically, along with infrastructure (roads, electricity and so on). The redeveloped area may well also be safer than the medina it replaced – certainly the roads are well-lit and police cars can easily patrol them.

And arguably Monastir attracts more tourists now that is has something more than 'just another medina' to offer, with most tourists making a beeline for the mausoleum.

Conclusions

For over a thousand years medinas had enormously significant implications across the Sahel. They were the seat of government, religious centres, the focus of economic activity as well as concentrations of learning and progress. What for such a long period of time was a symbol of modernity, however, has now become a symbol of tradition. The debate that continues is whether medinas have any significant implications any longer for the Sahel. Would it matter if all the medinas were excised in the way that Monastir's has been?

Chapter 14

Security in Monastir and the Sahel

Introduction

The purpose of this chapter is to look at security in the Sahel, with particular reference to Monastir. As a term, *security* is typically thought of as defence security, but with the existence of cruise missiles and nuclear weapons with extensive destructive powers the notion of defence security is often too limiting and so the term *security* can usefully encompass much wider issues of making the future secure. Food security, environmental security and economic security are contemporary aspects of security that are important to citizens.

This chapter begins by examining defence security in the Sahel, a characteristic of the region that is immediately obvious in the defensive nature of the medina walls in Sousse, Kairouan and Mahdia, and then develops ideas of Tunisian security in the global economy through investment in higher education, the film industry, tourism and the Internet.

Defence Security

As noted in the discussion of fishing, Tunisia has 1300 km of coastline and so has easy sea access. With the generally low-lying coastal plains this has meant that Tunisia is relatively accessible by sea and so security has always been an important issue for the peoples inhabiting the coastal areas. Defence security is evident within the place names in Tunisia. Small fortress towns originated from a combination of grain storage and stronghold. The name for such fortified grain storages was a *ksar*, a name that is now seen in place names such as Ksar Hellal south of Monastir. The plural of *ksar* is *ksour*, a name that is found in the town of Ksour Essaf south of Mahdia and indeed in the whole of the south of Tunisia which is called El Ksour.

Defence security is highly visible in all the main towns of the Sahel. The medinas of Sousse, Kairouan, Monastir, Mahdia and Sfax all have thick, high walls that are clearly intended to keep out unwanted visitors. Figure 14.1 shows the medina wall in Sfax, clearly a defence against outside attack. These defensive medina walls and their impressive gates are maintained and kept in very good condition, even today when they may constrain rather than enhance quality of life.

Figure 14.1   The medina wall and the gate *Bab Diwan* of Sfax (Walls and gates such as these are common features of medinas in the Tunisian Sahel)

One of the best defensive sites in the Sahel is Monastir. It is almost an island, separated from the rest of the Sahel by a sebkha and with only two main routes from the rest of the Sahel onto the Monastir promontory. There is evidence of habitation in the area since the 4th century BC, and Hannibal developed the town of Leptis Minus near Monastir in 203 BC. Julius Caesar established Ruspina, the forerunner of Monastir, in 46 BC as a coastal fortress for access to Africa Nova and for controlling Kairouan (see Chapter 10). One of the most dramatic buildings in Monastir is the ribat, built in 796 AD by El Aayoun and the first of the chain of forts and ribats that extended from Alexandria in Egypt to Ceuta in Morocco. As with the ribat in Sousse, the Monastir ribat is in part a fortress and in part a religious building (see Figure 14.2). The ribat of Monastir pre-dates the ribat of Sousse, and emphasises the defensive character of the Sahel region. Monastir thrived as a fortress town, with travellers in the 11th century AD noting the wealth of the town with its ribat, mosque, houses, baths, shops and a main square. The medina of Monastir developed both internal and external defences. Externally the medina wall is thick and high with originally up to 10 gates comparable to that shown in Figure 14.1 to control access. Internally, the medina was divided into three sections, each separated by a wall as thick and high as the external wall. The ribat made up the north east corner of the medina and was the obvious sign of the defensive nature of Monastir.

Figure 14.2   The ribat of Monastir

Monastir has been a trading centre for centuries. In the 18ᵗʰ century Monastir exported olive oil, soap and hats to Egypt and olive oil to Mecca and Medina. After the start of the French protectorate in 1881 Monastir became a garrison town run from Tunis, but was significant enough to house consulates of nine countries including the United Kingdom, the United States, Spain and Germany.

During World War II General Rommell used Monastir as a base for part of the retreat of the German forces heading northwards towards Tunis. While the major battles were in the interior, such as at the Kasserine pass, Monastir acted as a staging post for the planning of the retreat.

The New Security

Changing concepts of security

While defence security is clearly evident in the landscape of the Sahel, with ribats and strongly fortified medina walls, the concept of security needs to change in the 21ˢᵗ century. The thick, high walls of medinas are no longer sufficient to protect citizens from outside pressure. This section looks at the how the Sahel and Tunisia as a whole is positioning itself to enhance its security in the global economy. The accompanying box (Box 14.2) presents the government's view of the relationship between national needs and global issues, while this section examines four threads

of change concerned with Tunisian security in the global economy, namely higher
education, the film industry, tourism and the Internet.

Box 14.1 Conservation in Kairouan

In Kairouan the Association de Sauvegarde de la Medina (ASM) promotes the
conservation of architecture and Islamic values. The ASM successfully
achieved UNESCO World Heritage status for the medina of Kairouan in 1988,
nine years after the designation of the medina in Tunis. The ASM has
responsibility for the conservation of mosques, mausoleums, the facades of
buildings and the ancient wall of Kairouan medina. It employs up to 50 skilled
craftsmen and labourers to carry out the conservation. Up to November 2002
150 facades within the medina have been restored. The ASM also has an
advisory role on planning procedures in Kairouan. New buildings must not
exceed the height of the medina wall, and developments must be in line with
the traditional architecture of the medina. There are three main conservation
schemes in Kairouan medina.

The main north-south retail artery and adjacent souks have been subject to
restoration and conservation.

A *route touristique* has been established from the southern Bab el Chouhada to
the Grand Mosque in the north east corner of the medina, connecting several
important Islamic buildings (for example the Mosque des Trois Portes) along
the way.

A new scheme, funded by the World Bank, will link the two existing projects.
A new interpretation centre will be built opposite the Grand Mosque. A new
paved route will be constructed to encompass less obvious tourist attractions
and incorporate the more depressed and peripheral north east corner of the
medina in the broader cultural and economic process of rehabilitation.

*Adapted from the work of Ian Humphrey, UCL Geography undergraduate,
November 2002.*

Higher education

The number of young Tunisians in higher education has grown substantially since
independence. In 1966 only 2 per cent of the 19-24 age group participated in
education. By 1976 this figure had increased to 6 per cent and by the year 2000 to
19 per cent. Tunisia now has seven universities plus a new virtual university using
Internet access.

University of Manouba   University of 7 November at Carthage
University Ez-Zitouna, Tunis   University of the Centre
University of Tunis   University of Sfax for the South
University of Tunis El Manar   Virtual University

The University of the Centre is mainly based in Monastir and Sousse. In Monastir there are four faculties offering courses in science, medicine, dentistry and pharmacy plus high schools offering courses in engineering, science and biotechnology. Sousse has three faculties offering courses in medicine, law, economics, politics and social sciences. Other parts of the University of the Centre are located in Kairouan, Mahdia, Hammam Sousse, Chott Meriem and Moknine, and in total the university has 28 different faculties, institutes and schools. There is a strong emphasis in higher education in Tunisia on practical skills such as medicine and science, an indication of developing a knowledge base in students that is geared to further national development and improved national security.

Film industry

The outputs of the film industry in Tunisia are well known in the west. The films *Star Wars* and *The English Patient* were largely filmed in the south of Tunisia, using the extensive and imposing desert landscapes to provide location backgrounds. In Monastir the ribat has been the scene of several films, including *Jesus of Nazareth, Raiders of the lost Ark* and *Life of Brian*. It is somewhat ironic that a ribat is used, albeit with extensive modifications, to represent both a market in Jerusalem at the time of Christ and a place where a Roman centurion castigated a Jew for misbehaving.

The film industry has favoured Tunisia because of its good weather, high sunshine hours and low production costs. In turn Tunisia welcomes and promotes the international film industry because it provides high prestige, encouragement for tourists to visit Tunisia, and income. There is even a travel company (Tataouine Tours) specialising in visits to Star Wars film locations in the El Ksour region of Tunisia.

Tourism

The first major sun-and-sand tourist destination in Tunisia was established between Sousse and Monastir in the 1960s. Since then there has been a massive expansion in hotels in the area, and there are now 28 hotels in the 10 km strip of land between the two towns. This area was formerly called the Dhkila, an area favourable to growing early season vegetables because of its sandy soils and proximity to the sea. On the side of the Dhkila close to the sebkha an airport was built specifically to serve tourist flights. The airport, now called the Habib Bourguiba Airport, had 440,000 passenger movements in 1996, increasing to 880,000 passenger movements in 1999. In 1985 a new metro was built to link Sousse and Monastir, with stations in between to link the hotels with the two towns. As well as the Dhkila strip, dramatic hotel expansion has occurred north of Sousse, around the new marina of Port El Kantaoui, and also immediately north of Mahdia.

To support the expansion in hotel numbers schools for waiters and hotel managers have expanded, and there is direct employment in building, maintenance, electrical services, textiles and cafes, as well as greater demands for language skills. In the larger tourist hotels it is not uncommon to find waiters who can speak four foreign languages well, and have an ability to communicate in another six.

The tourist industry in Tunisia now has a revenue of over two billion dinars per annum, contributes six per cent of the country's Gross Domestic Product and employs about 300,000 people. Tourism in Tunisia is however in something of a quandary. On the one hand it is over specialised on a narrow sun-and-sand market that is not particularly wealthy. The average spend per tourist is only about 60 TD (approximately £30) per day (Hazbun 1997), including food and accommodation. The market is narrow, and even recent marketing campaigns to attract new visitors from eastern Europe have simply maintained the relatively narrow base with each tourist arriving with little spending money. On the other hand the country has cultural riches attractive to tourists, yet packaged in a very controlled way. Most tourists visiting the Sahel spend the majority of their time in the hotels on the coast and participate occasionally in organised day trips to the packaged major sites such as the Roman amphitheatre in El Djem, the grand mosque in Kairouan or the pretty but largely unremarkable village of Sidi bou Said on the Bay of Tunis. Such trips do not involve an overnight stay: the trips that do are to the south of Tunisia, typically packaged as two day desert safaris with visits to date palm oases, sand dunes and film locations in the Sahara desert. The combination of over specialisation and overt control has created a problem for Tunisian tourism: total revenue is limited because of the structure of the sector.

Revenues have been decreased by the fall in international tourism since 11 September 2001. The fall has hit Tunisian tourism particularly hard because it is a muslim country which drops it to the bottom of the list of country destinations when the World Trade Centre in New York and the Pentagon in Washington DC have been attacked in the name of Islam. Tunisia's image in this regard has not been well served by the attack by Al Qaeda on a synagogue on the island of Djerba in April 2002, and by a passenger of Tunisian origin (although a naturalised Swedish citizen) trying to take a hand gun on board a Ryanair plane in Stockholm in August 2002. Terrorism has hit the tourist industry in the Sahel very hard, with many hotels having low occupancy rates in late 2001 and in 2002.

As one local politician in Bekalta put it, tourists need to stick to the patterns arranged for them (the *cadre touristique*) and not venture into the other parts of Tunisia that are not on the tourist trail. However, tourism in other countries is changing and tourists becoming more involved with the countries they visit. Hazbun (1997) summarises the trend:

They [the tourists] are seeking sights to read, explore and interpret. Instead of generic vacations they want meaningful experiences. In this way there has been a re-integration of place into the tourist product.

Tourism in Tunisia goes against this trend. Tourists are encouraged to keep to the *cadre touristique* by the government and by tour operators, and these tourists have low spending power anyway. Hence the self-imposed revenue limits. Tunisia could both improve its contribution to its own security and expand its tourism revenues not by focusing on the supply side and providing more hotels, but by focusing on the demand side and providing a more liberal experience where the tourist and non-tourist segments of the economy merge as they do in the largest tourist destinations in the world, France and the United States.

Internet

President ben Ali announced in November 1997 an initiative to promote the use of the Internet in Tunisia. The national Internet strategy has three main elements.

Improvement in infrastructure
Development of a valid legal framework for e-commerce
Development of human resources, in particular in young people.

During the period of the Ninth Economic Development Plan (1997-2001) some 1.5 billion TD was invested in the Internet in Tunisia. At the start of the initiative in 1997 the Internet connection speed was 0.5 Mbits per sec and there were only about 100 users in the country. By the year 2000 the connection speed had increased to 15.5 Mbits per sec and 56 cyber-cafes out of a planned 110 had been opened. By August 2002 there were 461,000 Internet subscribers in Tunisia, with Internet access available in all universities, high schools and secondary schools. Tunisia has an agency dedicated to the Internet, the Agence Tunisienne d'Internet (ATI). The country now has 12 Internet Service Providers (ISPs), seven in the public sector to provide services to government departments and education, and five in the private sector for open access, such as PlaNet Tunisie and Tunet.

Many government functions are now available through the Internet, and enrolment in the universities is also provided by Internet links. The use of the Internet for business is being encouraged through the development of the e-dinar, and in 2005 Tunisia will host the World Summit on the Information Society.

Box 14.2 Tunisia and the global system

Interview by United Press International with President ben Ali
Tunisia News 15 July 2000

UPI: How do you perceive the relationship between national needs and requirements and the growing globalization of the international system? What are the effects of globalization on national identities?

President ben Ali: With the opportunities brought about by globalization, new challenges have emerged. Whereas a number of nations have benefited from globalization, the circumstances of a number of other nations have grown in complexity and the developmental gap between and within nations has widened. We, in Tunisia, have chosen to avail ourselves of the opportunities that globalization offers and that can be exploited.

Continued

Our country which has consciously and responsibly enlisted in the course is aware that globalization, while opening wide perspectives for development and progress, generates negative effects in the social field. Accordingly, we have based our development strategy on the coincidence between the economic and social dimensions, so as to avoid social rifts that could damage the stability and balance of our nation.

In compliance with this vision, we have developed a comprehensive strategy to meet the challenge of competition that is one of the features of the dynamics of a globalized economy. In the development process, we turned our attention to productive structures, to technological progress, and to the reinforcement of competitiveness in all fields. That is to say we established a more varied economy and, to do so, we adopted a strategy for a comprehensive upgrading of our enterprises, a strategy which we developed successfully and to a high degree. We opened new scopes for our exports and we set up mechanisms and incentives to develop our export potential. We provided for appropriate conditions to attract direct foreign investments so as to provide for an increased participation of the private sector in the investment effort, notably in the high added value sectors and in the sector of modern technologies.

Based on our awareness that investment in knowledge is the only way to meet the challenge of competitiveness, we have striven to increase the financial resources devoted to scientific and technological research, to reach 1 per cent of our gross domestic product in 2004; we have also encouraged partnership with advanced countries in this field.

We have also seen to the adequate preparation of our human resources. Our young are thus provided with training in future-oriented skills so that they may be ushered into the post-industrial era: an era based on immaterial economy in which the only important elements are intelligence and the competence of human resources.

In this respect, we have endeavoured to spread the teaching of computer science and expanded the Internet to schools, universities, homes, as well as various public locations so as to draw maximum benefits from the data and services provided by this network and ensure that our young are increasingly open on a world where knowledge is constantly being renewed.

We have also provided for our universities and research centres to be linked with major scientific, technical, and cultural data banks worldwide. We have increased our links with world data networks as an interface with advanced economies worldwide.

Continued

Human and economic development indicators show that Tunisia was successful in achieving a difficult equation in meeting national needs and complying with the requirements of globalization. Today, our economy is considered the first in Africa in terms of competitiveness, according, among other sources, to the recent report issued by the World Economic Forum and the Harvard Institute for International Development.

As regards the second half of your question, we believe that opening up on the regional and international environment and reinforcing partnership as well as positive interaction with changes and developments on the international scene should not keep us from preserving the basic elements of authenticity and national culture. We believe that cultural specificities are not incompatible with the requirements of globalization.

On the contrary, we consider that cultural diversity and civilizational variety enhance universal values that need to be permanent for all nations and within both national and international policies.

Tunisia, though a small country, geographically speaking, is a great country with regard to its cultural and historical heritage and its ambition to influence and shine on the world.

Considered for more than 3000 years as the home of the culture of diversity, Tunisia has always called for a dialogue between civilizations and endeavoured to steer mankind clear of the hazards of conflicts between nationalities and religions. In our interaction with globalization, we have based ourselves on our country's great civilizational heritage as well as on an approach considering globalization as a new opportunity offered to our country to be more present on the international scene.

In this regard, culture in our country has become a major area of activity and production to root national identity within a world where the disappearance of borders threatens the foundations of national character.

To reach this goal, we have encouraged innovation in all artistic and cultural fields and expressed our commitment to gradually increase budget allocations devoted to culture so that they may be doubled and reach 1 per cent of the state budget in 2004. The culture we are ambitious to achieve is a culture that benefits from all the factors of today's progress and contributes to the dissemination, openness, and tolerance that is to say a culture that, to a certain extent, guards our character and is capable of highlighting our contributions to universal culture.

Chapter 15

The Tunisian Sahel in Context

Overview

Taken together, the chapters in this book provide a fairly comprehensive overview of the changing geography of the Tunisian Sahel. Even modestly stated this is a bold claim, and some reservations need to be acknowledged. We have relied fairly exclusively on information sources written in English – including academic publications, unpublished reports, various data sources, newspapers and the internet; as well as field observations and student reports. This book would have benefited from a survey of French – and perhaps more interestingly Arabic – sources, as it would have from the insight of Tunisian students. There are certain important topics we have omitted or mentioned only in passing because of a lack of accessible literature and data – for example the Berbers, Gulf state migration and the informal economy. Other omissions perhaps reflect our biases and our objectives for the field course from which this book project has emerged – neither of us for example feels confident writing in depth about Tunisian culture – art, for example, or food or music. Finally, we are both acutely aware that we are 'outsiders' writing about a country where we have only been visitors (if fairly frequent visitors over a fairly long period of time). We have assiduously tried to avoid making any claims of 'authenticity' or writing about themes that require an 'insider's' insight.

The Changing Geography of the Tunisian Sahel

Even given these limitations, we hope that an overview has emerged. Rather than revisiting the detail of each of the preceding chapters, it is worth emphasising three central themes that characterise the changing geography of the Tunisian Sahel. One is the idea of continuity and change. This may be a fairly hackneyed and tired phrase, but we feel that it captures perfectly Tunisia's dynamic character. The landscape, architecture, culture and society today bear the imprints of the astonishingly wide array of influences on Tunisia, from the Phoenicians, through the Romans and Arabs to the French. It is worth pausing for a moment to reflect that in the history of the world these are by no means inconsequential influences, and few other countries can have had such direct contact with so many formative civilisations. At the same time, the visitor to Tunisia is struck by the forward momentum of the country, from progressive economic reforms and agricultural development through a modern, liberal interpretation of Islam to its people's entrepreneurial spirit.

A second overarching theme is the idea of Tunisia as being 'on the edge' or 'at the margins'. As explained in the book's Introduction, this is of course a play on the meaning of the Arabic term Sahel. And indeed Tunisia is in many ways at the margins of both Europe to the north and Africa to the south – its physical environment and climate, its demographic trajectory and its economic and political profile all fall 'in between' those to the north and south. It has relatively few clear links to the south, while the European Union continues to hold it at arm's length. A clear distinction that has emerged, however, and that needs to be emphasised here, is that while Tunisia may be located 'at the margins' in all these ways, it is by no means *marginal*. Indeed quite the opposite – we have seen how on scores of economic development and political probity it is close to the world average; it is largely stable and safe rather than extreme, and it figures centrally on at least certain EU agendas. It may not share the same international spotlight as its neighbours Libya and Algeria, nor does it spring immediately to mind as part of the Middle East – but given the reasons why these other countries generally do attract attention today its relative invisibility is probably to the credit of Tunisia.

A third recurrent theme has been the tension – sometimes positive and sometimes negative – between tradition and modernity. In the coverage in this book this tension has been most overt in the varying responses to the challenges of the Sahelian medinas. To a lesser extent it has emerged as a theme too in respect of the role of women in Tunisia, and in the political significance of Islam in Tunisia. In each of these latter arenas Tunisia appears to have resolved the tension between tradition and modernity with great success. Once again Tunisia stands out precisely because this tension does not stand out. While France continues to debate whether or not Muslim girls should wear headscarves at school, Tunisia has a long-established secular education system. While even an understanding of the cultural context cannot disguise the parlous status of women in many Islamic countries, women in Tunisia enjoy rights that are comparable to those in any secular, democratic state in the world. And while most countries in the Arab world are fragmenting as a result of the politicisation of Islam (or the Islamicisation of politics), Tunisia seems to have achieved a genuinely secular state in a genuinely Muslim nation.

Tunisia in the Arab World

One of the reasons for emphasising broad themes rather than specific details in the preceding overview has been to try to place the Tunisian Sahel – and indeed this book as a whole – in a wider context. Another way to do this is to shift from the thematic context to the regional context. Chapters in the book have already explained why Tunisia does not clearly fit in either the European or the African context; nor as we saw in the Introduction is it usually considered part of the Middle East. Instead, the most obvious regional context for the study of Tunisia is either as part of the Maghreb region of North Africa, or more widely as part of the Arab world (Findlay, 1994).

Different authors have adopted different definitions of the Arab world, both in terms of geographical reach and characteristics. Geographically it is normally taken to include the Arabian peninsula and North Africa including the Maghreb. It is overwhelmingly Muslim – although of course the 'Muslim world' extends far beyond the Arab world (to Nigeria and Indonesia, for example). It shares a common language (Arabic) although there are important regional dialects; and a common culture, although again with significant regional differences. Furthermore, Tunisia clearly has economic and political links with the Arab world – for example through trade, labour migration and membership of the Arab League. On the other hand, and as alluded to in the previous section, Tunisia is in several important ways distinct from most of the rest of the Arab world. Its political and education systems are broadly secular, women enjoy an unrivalled status and it is certainly more oriented towards Europe than most other Arab countries – at least in part because of a colonial past.

There is less ambivalence about the Maghrebian context. The countries of North Africa share comparable historical antecedents and colonial pasts. They share a similar physical environment and climatic conditions, demographic characteristics and broadly comparable economic indicators. They are increasingly viewed as a single political entity by the European Union. And this is all above and beyond a shared religion, language and culture. Even within this context, however, Tunisia stands out for its stability, secularism and openness.

Teaching and Learning in the Field

A final way we are keen to place Tunisia in a wider context is to present it as a 'case study' for a fieldwork site. Our assertion here is not necessarily that fieldwork in Tunisia will teach students about other countries in the Maghreb or Arab world, although it will certainly expose them to an environment which is largely common to them all. Instead it is a more pedagogical assertion, namely that fieldwork is the best way to teach and learn about another country and context.

It is worth reiterating that this book emerged from field courses on the Tunisian Sahel taught first at Durham then at UCL. For all the reasons illustrated in this book, the Sahel is a fascinating region for the geographer, so how can the geographer interact with the Sahel through direct field observation? This final section provides a summary of how we have done it since 1999.

The Basic Model

The basic teaching model we have used is to separate the field lectures and the student investigations into morning and afternoon sessions. The morning periods are taken up with lectures delivered in the field on the main topic of the day. The delivery of the lectures in the place of study, for example in mosques, fishing ports or agricultural fields (Figure 15.1), adds substantially to the student learning experience. The afternoon periods are typically taken up with student group exercises to collect information to develop the themes examined in the morning session. The collection of information first-hand is a prized asset of fieldwork. We often use the dictum of Carl Sauer that: 'the mode of locomotion should be slow, the slower the better, and be often interrupted by leisurely halts to sit on vantage points and stop at question marks' (Sauer, 1956).

Figure 15.1 A field lecture on security in the ribat in Monastir

Fieldwork is followed up in the evening with discussions of the issues raised during the day. These evening discussions give students the opportunity to test out their ideas with their peer group, to report on their findings in a structured way, to compare reactions to the same or different places and to consolidate their learning in the lectures through real field experience.

Urban Areas

Urban areas are convenient. In logistical terms, buses, trains and taxis go from city to city, town to town and village to village. In the case of the Tunisian Sahel the region can be characterised as one of large, connected villages and towns, so the study of urban areas is an essential part of fieldwork there. Table 15.1 shows the pattern of a typical study day in the Sousse medina. The objectives of the day's fieldwork are to provide an orientation to Sousse, to introduce the religion of Islam and to investigate continuity and change in the form and function of the medina. The lectures in the morning are concerned with Islam and with the form and function of the Sousse medina, and by extension with the form and function of the majority of medinas in the Arab world.

Table 15.1 The timetable for field work in the Sousse medina

Time	Activity
0830	Walk to Sousse medina via the tourist and French sectors
0900	Lecture on Islam: Sousse grand mosque
1000	Lecture on security in the Maghreb: Sousse ribat
1030	Group walking transects of the Sousse medina: kasbah, medina walls, fish, meat and vegetable markets, residential areas, religious buildings
1200	Lecture on the form and function of the Sousse medina
	Briefing on the medina information collection exercises
1300	Lunch break
1330	Information collection and investigations in small groups to answer the questions shown in table 15.2
1630	Meet in central Sousse
	Return to accommodation
1930	Group discussions of medina form and function based on the afternoon investigations
2130	Close

Table 15.2 Questions to guide student information collection in the Sousse medina

Theme	Questions
Urban form	Is urban renewal taking place in the medina?
	What are the most obvious planning problems in the medina?
	What are the types of solution to the planning problems?
Urban function	What are the medina's prime functions?
	How important is tourism?
	To what extent do the selling activities in the medina show links with the global economy?
Urban space	What factors have created the morphology of the medina?
	How far are activities in the medina concentrated spatially?
	Is there a contest for space in the medina?
	How is it resolved?
Cultural expression	Where are the women in the medina?
	What are they doing?
	How does the architecture in the medina reflect Islamic culture?
	How different is the structure of the medina from European town centres?
Security	What is the function of the wall of the medina?
	For whom does the medina wall make no difference?
	What are the security issues facing the medina and its inhabitants?

The afternoon investigations are designed to enable students to collect information first-hand about the Sousse medina. Table 15.2 lists five sets of questions about the Sousse medina: a group of about six students is allocated to each set of questions and invited to collect information to answer them. The evening discussion uses the five sets of questions as its agenda, and the information collected by students nourishes the discussion at the same time as emphasising and illustrating the main points from the lectures.

Rural Areas

Rural areas present something of a challenge to fieldwork. A group of foreign students is always noticeable in any environment, but in rural areas the group is particularly noticeable. Rural areas often present problems of access, although the rewards of direct contact with people and with landscapes can be high.

We have used the same division of morning lectures and afternoon investigations as in urban areas. In the afternoon students are divided into groups and each group is allocated a transect to illustrate the main themes of the day's fieldwork. Table 15.3 lists 11 questions that students seek to answer during the afternoon transect session by direct field observation. We have worked with these questions in the areas of irrigated agriculture around Bekalta, Sahline and Sidi bou Ali. The information collected in response to these questions is then consolidated for the evening discussions.

The agenda for these discussions is cross-cutting in nature, in that it tries to draw out some of the major themes of rural development, illuminated by the information collected through the questions in Table 15.3. The following agenda for the evening discussion of rural areas has proved useful:

What is the impact of the physical environment on agriculture and rural development in the Tunisian Sahel?
Is agriculture in the study region sustainable?
How are rural products marketed?
What is the basis for food security in the Sahel?

Table 15.3 Questions to guide the collection of field evidence in rural areas

What are the activities of the men you see during your transect?
What are the activities of the women you see during your transect?
What are the crop types in the areas of irrigated agriculture?
What are the crop types in the areas of non-irrigated agriculture?
What are the main methods of irrigation?
What happens in the areas with no irrigation?
What are the types of agricultural investments, e.g. irrigation, plastic cloches, electricity, water pumps, tractors and other machinery?
What is the quality of the infrastructure investments?
How is water lost?
What are the elements of continuity in the landscape?
What are the elements of change in the landscape?

Conclusion

A casual reader of this book might conclude that we have an unabashed, unashamed enthusiasm for Tunisia. He or she would be wrong. We trust we have been even-handed – there are aspects of Tunisia that deserve to be celebrated, there are others that deserve roundly to be criticised. Many steps have been taken towards achieving gender equality – but more are needed. The political system may be secular and democratic, but human rights violations sully that reputation. Christians and Jews and their beliefs and practices are happily welcomed in Tunisia – but the Berbers are shockingly marginalized. Tourism in Tunisia makes a significant contribution to the economy, but the sector is self-limiting because of its structure and the tourists are not encouraged to stray away from the set routes. Tunisia continues to battle with water supplies in the search for sustainable development. And it is worth reminding the reader that the main reason successive field classes from Durham and UCL have visited Tunisia is because it is cheap to get there, it is not far away and it is safe.

Our real enthusiasm for Tunisia is as a geographical 'laboratory'. Put simply it is a wonderful place to practice geography. It has a varied physical environment and climate, a rich history, and a culture and religion completely different from those to which students from the industrialised world are really accustomed. And what makes Tunisia unusual is that all of these aspects of its geography are so readily accessible to the scholar, student and tourist too. This allows all of us to see geography in practice, to see change in continuity and continuity in change.

References

Abu-Lughod, J. (1984) 'Culture, modes of production and the changing nature of cities in the Arab world', in J. Agnew, J. Mercer and D. Sopher (eds) *The City in Cultural Context*, Boston: Allen and Unwin, 94-119.

African Development Bank (Annual) *African Development Report*, Oxford: Oxford University Press.

Alagiah G (2002) *A Passage to Africa*, London: Time Warner Publications.

Al-Bayati, B. (1984) *The City and the Mosque*, Oxford: AARP.

Anon (2002) *The Religion of Islam*, Islamic Affairs Department, The Royal Embassy of Saudi Arabia in Washington DC, http://www.iad.org/.

Barry, R.G. and R.J. Chorley (1998) *Atmosphere, Weather and Climate*, London: Routledge, 7th edition.

Beaumont, P., Blake, G., Wagstaff, J.M. (1988) *The Middle East: A Geographical Study*, London: Fulton.

Berry-Chikhaoui, I. (1998) 'The invisible economy at the edges of the medina of Tunis', in R.A Lobban, Jr (ed.) *Middle Eastern Women and the Invisible Economy*, Gainesville: University of Florida Press, pp.215-30.

Bianca, S. (2000) *Urban Form in the Arab World: Past and Present*, London: Thames and Hudson.

Blake, G. and R. Lawless (eds) (1980) *The Changing Middle Eastern City*, London: Harper and Row.

Borowiec, A. (1998) *Modern Tunisia: A Democratic Apprenticeship*, Westport: Praeger.

Brett, M. and Fentress, E. (1996) *The Berbers*, Oxford: Blackwell.

Brierley, G.J. (1981) 'The wadis of the Sousse region, central Tunisia', in R. Harris and R.I. Lawless (eds) *Field Studies in Tunisia*, University of Durham, 4-17.

Bucaille, M. (1976) *The Bible, The Qur'an and Science*, Paris: Publishers Seghers, translated from the French by A. D. Pannell and M. Bucaille.

Cassarino, J-P. (2000) *Tunisian New Entrepreneurs and their Past Experiences*, London: Ashgate.

Chazan, N. et al. (1988) *Politics and Society in Contemporary Africa*, Boulder: Lynne Rienner

CIA (2002) *The World Factbook 2000: Tunisia*, http://www.cia.gov/cia/publications/factbook/ geos/ts.html, 17 June 2003.

Clancy-Smith, J.A. (1994) *Rebel and Saint*, London: University of California Press.

Clarke, J. and D. Noin (eds) (1998) *Population and Environment in Arid Regions*, Paris: UNESCO.

Clarke, J. and H. Bowen-Jones (eds) (1981) *Change and Development in the Middle East*, London: Methuen.

Climatic Research Unit (2002) 'Tyndall Centre for Climate Change Research', data set TYN CY 1.0, available at http://www.cru.uea.ac.uk/, 1 September 2002.

Cornelius, W.A., Hollifield, J. and Martin, D. (1994) *Controlling Immigration*, Stanford: Stanford University Press.

Dalacoura, K. (1998) *Islam, Liberalism and Human Rights*, London: IB Taurus.

Davidson, B. (1991) *Africa in History*, 3rd Edition, London: Phoenix.

Davies, S. (1981) 'The Nebhana irrigation scheme: an instrument of rural development', in R. Harris and R.I. Lawless (eds) *Field Studies in Tunisia*, University of Durham, 18-28.

de Montequin, F. (1983) 'The essence of urban existence in the world of Islam', in A Germen (ed.) *Islamic Architecture and Urbanism*, Damman: King Faisal University, 43-58.

Despois, J. (1955) *La Tunisie Orientale – Sahel et Basse Steppe – Etude Geographique*, Paris: Presses Universitaires de France.

Drysdale, A. and Blake, G. (1985) *The Middle East and North Africa*, Oxford: Oxford University Press.

Europa (2003) *The Middle East and North Africa 2003*, London: Europa Publications.

European Commission (1976) *Telex Mediterranean 29 April 1976*, Brussels: European Commission.

Fage, J. (2002) *A History of Africa*, 4th Edition, London: Routledge.

Ferchiou, S. (1998) '"Invisible" work at home: the condition of Tunisian women', in R.A Lobban, Jr (ed.) *Middle Eastern Women and the Invisible Economy*, Gainesville: University of Florida Press, pp.187-214.

Field, M. (1994) *Inside the Arab world*, London: John Murray.

Findlay, A.M. (1994) *The Arab World*, London: Routledge.

Findlay, A.M. and Paddison, R. (1984) *Planning the Arab City*, Oxford: Pergamon.

Fisher, W.B. (1978) *The Middle East*, London: Methuen, 7th edition.

Fishman, M. and Khan, H-A. (1994) (eds) *The Mosque: History, Architecture and Regional Diversity*, London: Thames and Hudson.

Gerner, D.J. and Schrodt, P. (2000) 'Middle Eastern politics', in D.J. Gerner (ed.) *Understanding the Contemporary Middle East*, London: Lynne Rienner, pp.81-128.

Giammusso, M. (1999) 'Civil society initiatives and prospects of economic development: the Euro-Mediterranean decentralized cooperation networks', *Mediterranean Politics* 4 (1) 25-52.

Gillespie, R. (1997) 'The Euro-Mediterranean partnership', *Mediterranean Politics* 2 (1) 1-5.

Gleick, P.H. (1992) 'Effects of climate change on shared fresh water resources', in I.M. Mintzer (ed.) *Confronting climate change: risks, implications and responses*, Cambridge: Cambridge University Press.

Harris, R. and R.I. Lawless (1981) *Field Studies in Tunisia*, University of Durham.

Hattstein, M. and P. Delius (2000) (eds) *Islam: Art and Architecture*, Cologne: Könemann.

Hazbun, W.A. (1997) *The commodity of tourism in Tunisia*, presented at the Centre d'Etudes Maghrébines à Tunis (CEMAT), 13 June 1997.

Hillenbrand, R. (1983) 'Some observations of the use of space in medieval Islamic buildings', in A. Germen (ed.) *Islamic Architecture and Urbanism*, Damman: King Faisal University, 17-28.

Hillenbrand, R. (1994) *Islamic Architecture: Form, Function and Meaning*, Edinburgh: Edinburgh University Press.

Hollis, G.E. and M.R. Kallel (1986) 'Modelling natural and man-induced hydrological changes on Sebkef Kelbia, Tunisia', *Transactions Institute of British Geographers* NS 11(1), 86-104.

Hopwood, D. (1992) *Habib Bourguiba of Tunisia*, London: Macmillan.

Hutt, A. (1977) *Islamic Architecture in North Africa*, London: Scorpion.

INS (2000) *Statistiques Economiques et Sociales de la Tunisie, Commerce extérieur par pays*, Institut Nationale de la Statistique, Ministry of Development and International Cooperation, Tunis, http://www.ins.nat.tn/.

INS (2002) *Statistiques Economiques et Sociales de la Tunisie, Données Economiques et Financières*, Institut Nationale de la Statistique, Ministry of Development and International Cooperation, Tunis, http://www.ins.nat.tn/.

IPCC (2001) *Climate Change 2001: Synthesis Report. A Contribution of Working Groups I, II, and III to the Third Assessment Report of the Integovernmental Panel on Climate Change* [Watson, R.T. and the Core Writing Team (ed.)] Cambridge: Cambridge University Press.

Izzi Dien, M. (2000) *The Environmental Dimensions of Islam*, Trowbridge: Redwood Books.

Jones, A. (1997) 'The European Union's Mediterranean policy: from pragmatism to partnership', in King R et al. *The Mediterranean: Environment and Society* London: Arnold, 155-163.

Ki-Zerbo, J. et al. (1980-90) *UNESCO General History of Africa*, 8 volumes, Paris: UNESCO.

King, R., Proudfoot, L., Smith, B. (eds) (1997) *The Mediterranean. Environment and Society*, London: Arnold.

Koser, K. and Lutz, H. (1998), 'The new migration in Europe', in K. Koser and H. Lutz (eds) *The New Migration in Europe*, London: Macmillan, pp.1-17.

Kuhn, T. (1993) *The Structure of Scientific Revolutions*, Chicago: University of Chicago Press, 3rd edition.

Lambert, J. (1971) 'The Cheshire cat and the pond: EEC and the Mediterranean area', *Journal of Common Market Studies* (10) 37-46.

Lawless, R. (1984) *The Middle Eastern Village*, London: Croom Helm.

Lawless, R. and Findlay, A.M. (eds) (1984) *North Africa*, London: Croom Helm.

LeVine, V. and Luke, T. (1979) *The Arab-African Connection*, Boulder: Westview.

Levitt, P. (1988) 'Social remittances: migration driven local-level forms of cultural diffusion', *International Migration Review*, 31(2), 411-47.

Licari, J. (1998) 'The Euro-Mediterranean partnership: economic and financial aspects', *Mediterranean Politics* 3 (3) 1-20.

Long and Reich (eds) (2002) *The Government and Politics of the Middle East and North Africa*, Boulder: Westview Press.

Ludlow, P. (1994) (ed.) *Europe and the Mediterranean*, London: Brassey's.

Marks, J. (1996) 'High hopes and low motives: the new Euro-Mediterranean partnership initiative', *Mediterranean Politics* 1 (1) 1-24.

Marks, J. and M. Ford (2001) *Tunisia: Stability and Growth in the New Millenium*, London: Euromoney.

Mazrui, A. (1986) *The Africans*, London: BBC.

Mbembe, A. (2002) 'At the edge of the world', in M. Beissinger and C. Young (eds) *Beyond State Crisis: Postcolonial Africa and Post Soviet Eurasia in Comparative Perspective*, Washington: Woodrow Wilson Center Press, pp.53-80.

McGuiness, J. (ed) (2002) *Tunisia Handbook: The Travel Guide*, Footprint Handbooks.

Milton-Edwards, B. (2000) *Contemporary Politics in the Middle East*, Oxford: Blackwell.

Mitchell, T.D., Hulme, M., New, M. (2002) 'Climate data for political areas', *Area* 34, 109-112.

Moghadam, V. (2000) 'Population growth, urbanization and unemployment', in D. J. Gerner (eds) *Understanding the Contemporary Middle East*, London: Lynne Reiner, 239-62.

Monar, J. (1998) 'Institutional constraints of the European Union's Mediterranean policy', *Mediterranean Politics* 3 (2) 39-60.

Morris, P. and D. Jacobs (1998) *Tunisia, The Rough Guide*, London: Rough Guides Limited.

Murphy, E.C. (1999) *Economic and Political Change in Tunisia: From Bourguiba to Ben Ali*, Basingstoke: Macmillan.

National Research Council (1999) *Our Common Journey. A Transition Towards Sustainability*, National Academy Press, Washington DC.

New, M., Hulme, M. and P Jones (1999) 'Representing twentieth-century space-time climate variability. Part I: Development of a 1961-1990 mean monthly terrestrial climatology', *Journal of Climate* 12, 829-256.

O'Connor, A. (1991) *Poverty in Africa*, London: Belhaven.

Oliver, R. (1999) *The African Experience*, 2nd Edition, London: Weidenfeld and Nicolson.

Perkins, W. (1986) *Tunisia: Crossroads of the Islamic and European Worlds*, Boulder: Westview.

Piening, C. (1997) *Global Europe: The European Union in World Affairs*, Boulder: Lynne Rienner.

Prospero, J.M. (1999) 'Long-range transport of mineral dust in the global environment: impact of African dust on the environment of the southeastern United States', *Proceedings National Academy of Sciences* 96, 3396-3403.

Radwan, S., Jamal, V., Ghosh, A. (1991) *Tunisia: Rural Labour and Structural Transformation*, London: Routledge.

Reader, J. (1997) *Africa: A Biography of the Continent*, London: Hamish Hamilton.

Rivlin, P. (2001) *Economic Policy and Performance in the Arab World*, Boulder: Lynne Reinner.

Rogerson, B. (1998) *A Traveller's History of North Africa*, Gloucestershire: The Windrush Press.

Romeo, I. (1998) 'The European Union and North Africa: keeping the Mediterranean "safe" for Europe', *Mediterranean Politics* 3 (2) 21-38.

Rosander, E. and Westerlund, D. (eds) (1997) *African Islam and Islam in Africa*, London: Hurst.

Salem, N. (1984) *Habib Bourguiba, Islam and the Creation of Tunisia*, London: Croom Helm.

Salem-Murdock, M. and Horowitz, M. (eds) (1990) *Anthropology and Development in North Africa and the Middle East*, Boulder: Westview Press.

Sauer, C.O. (1956) 'The education of a geographer', *Annals of the Association of American Geographers*, 46(3), 287-99.

Shlaim, A. and Yannopoulos, G.N. (1976) *The EEC and the Mediterranean Countries*, Cambridge: Cambridge University Press.

Slim, H. (1996) *El Jem. L'antique Thysdrus*, Tunis: ALIF – Les Editions de la Méditerranée.

Stannard, D. (ed.) (1991) *Insight Guides: Tunisia*, Singapore: APA Publications.

Stone, R. and Simmons, J. (eds) (1977) *Change in Tunisia*, New York: University of New York Press.

Sutton, K., (1999) 'Demographic transition in the Maghreb', *Geography*, 84(2): 111-118.

Tovias, A. (1997) 'The economic impact of the Euro-Mediterranean free trade area on Mediterranean non-member countries', *Mediterranean Politics* 2 (1) 113-128.

Toyn, R. and G. Brierley (1981) 'An examination of the form and function of Tunisian urban centres', in R. Harris and R. I. Lawless (eds) *Field Studies in Tunisia*, 57-63.

Tsoukalis, L. (1977) 'The EEC and the Mediterranean: is "global" policy a misnomer?', *International Affairs* (54) 437-451.

Tunisia Online (2003) *Economy*, http://www.tunisiaonline.com/, 17 June 2003.

UNDP (Annual) *Human Development Report*, New York: Oxford University Press.

UNICEF (Annual) *State of the World's Children*, New York: UNICEF.

UTAP (2002) Tunisian Union of Agriculture and Fishing, Fishing in Tunisia, http://www.utap.org.tn/htmlang/utap/utap_p.htm, 20 November 2002.

White, G.W., Entelis, J.P., Tessler, M.A. (2002) 'Republic of Tunisia', in D.F. Long and B. Reich (eds) *The Government and Politics of the Middle East and North Africa*, Boulder: Westview Press.

Willett, D. (ed.) (2001) *Lonely Planet: Tunisia*, London: Lonely Planet.

World Bank (2002a) *World Development Indicators*, The World Bank Group, http://www.worldbank.org/data/wdi2002/, 17 June 2003.

World Bank (2002b) *Tunisia at a glance*, http://www.tunisiadaily.com/, 17 June 2003.

Xenakis, D.K and Chryssochoou, D.N (2001) *The Emerging Euro-Mediterranean System*, Manchester: Manchester University Press.

Zartman, I.W. (ed) (1991) *Tunisia: The Political Economy of Reform*, Boulder: Lynne Rienner.

Zussman, M. (1992) *Development and Disenchantment in Rural Tunisia*, Oxford: Westview.

Index

Africa 1, 3-4, 28, 51, 74-87
African Union 85
age structure 52, 82
agriculture 43, 48, 49-50, 62, 67, 70, 95-6, 98-104, 106, 146
aid 70-71
AIDS 79
air transport 83-4, 135-6
amphitheatres 93-4, 96
Arab culture, world 28-30
Arab League 85
architecture 114, 115, 123-4, 126

Barcelona process 72
ben Ali, Zine 38-41
Berbers 25
birth rate 54
Bourguiba, Habib 32, 33, 34-8

Carthage 25-7, 93
Changement, le 37-8, 41
child mortality 78, 79, 80
climate 14-20
Code of Personal Status 60-61
colonial rule 30-33, 56, 77, 125

dams 20-1, 100-101
death rate 54, 81
defence 131-3
demography 51-5, 81-2
development plans 43, 48-9
diplomatic links 84-5

economy 36-7, 41, 42-50, 67-73, 79, 136-7
education 61, 80, 135
El Djem 91-4
emigration 54, 55, 83
environment 13-24, 113
Europe 2, 44, 47-8, 51, 55-8, 66-73, 83, 84
Euro-Mediterranean Agreement 72
exports 43-4, 69-70, 79, 82-3, 99, 105

fasting 35
fertility 54, 81

film industry 135
fishing 104-106
Five Pillars of Islam 109-112
France 30-31, 43, 44, 47, 55, 56, 57, 67

gas 45-6
gender 35, 52, 57, 60-65
Germany 43, 44, 47, 55, 56, 57, 58
government 22-3, 34-41, 64, 82, 106
Gross Domestic Product 42-3, 78, 79

Hannibal 26
history 25-41, 91-7, 107-109, 120, 121, 132-3
human rights 39, 40

Ibn Battuta 113-4
Ibn Khaldun 29, 113
immigration 31, 83
imports 43-4
infant mortality 78, 79, 80
Internet 137
investment 44-5, 79
irrigation 19, 100-4
Islam 22, 29, 64-5, 76, 77-78, 107-120, 121
Islamism 37, 40
Italy 31, 33, 44, 47, 55, 56, 57

Kairouan 29-30, 118-120, 127, 134
Koran 22, 108-9, 111

Libya 44, 47, 58, 85
life expectancy 52, 78, 79, 80
literacy 78

Maghreb 1, 143
Mahdia 104, 106
marginality 1-4, 141-2
medinas 115, 121-30, 134, 141-2, 145
Mediterranean region 14-5, 25, 66-73
Middle East 3, 55, 75, 83, 84, 142
migration 31, 54, 55-9, 83
Monastir 32, 34, 38, 128-30, 132-3, 135
mosques 29, 115-20, 122, 127

Niger Republic 79
North Africa 47, 74, 75, 84, 86

oil 45
olives 95-6, 98-9

Phoenicians 25-6
politics 32-3, 35-6, 37-8, 63, 64
population distribution 53
population growth 51-5, 82-2
poverty 77

Qur'an see Koran

rainfall 16, 18-9, 20
Ramadan 35
religion 77-8, 107-120
remittances 58-9
ribat 116-7, 132, 133
Romans 26-8, 91-7

school enrolments 61, 80
sebkha 21

security 131-40
Sfax 53, 127, 132
soils 20-1, 24
Sousse 50, 53, 116-8, 123-7, 135-6
sub-Saharan Africa 74-87

Thysdrus 91-4
topography 13
tourism 46-8, 50, 83, 97, 127, 135-7
trade 43-4, 69-70, 71, 79, 82-3, 99, 105
Tunis 53, 84-5

universities 135

vegetation 21

walls 122, 126, 131-2
water 18-19, 22-3, 100-104
wheat 92, 95-6
women 35, 54, 57, 60-65

Printed and bound by CPI Group (UK) Ltd, Croydon, CR0 4YY

22/10/2024

01777626-0013